Science and Drama: Contemporary and Creative Approaches to Teaching and Learning

Peta J White · Jo Raphael · Kitty van Cuylenburg
Editors

Science and Drama: Contemporary and Creative Approaches to Teaching and Learning

Springer

Editors
Peta J White
School of Education
Deakin University
Burwood, VIC, Australia

Jo Raphael
School of Education
Deakin University
Burwood, VIC, Australia

Kitty van Cuylenburg
School of Education
Deakin University
Burwood, VIC, Australia

ISBN 978-3-030-84400-4 ISBN 978-3-030-84401-1 (eBook)
https://doi.org/10.1007/978-3-030-84401-1

This Springer imprint is published by the registered company Springer Nature Switzerland AG
The registered company address is: Gewerbestrasse 11, 6330 Cham, Switzerland

Foreword

Science and drama are natural bedfellows in education, it has always seemed to me. As the editors of this timely and diverse book remind us at the start, both 'are ways that we experience and make sense of the world'. Both have their roots in the same basic human instincts—of curiosity, to discover the world and ourselves in it through our senses and perceptions, and to try and understand what we perceive, and then to manage or control it. They have their common first manifestation in children's play, which is simultaneously scientific *and* artistic:

- Scientific because in play young children are using their senses to explore and map their world, their imaginations to make hypotheses and their developing faculties to test those hypotheses by trial and error, experiment and improvement, by using materials and making *models* of how the world beyond them works;
- Artistic because in play they are using their senses and faculties to imitate, express and explore their relationship with their surroundings and their feelings, giving aesthetic order and meaning to both their inner and outer worlds:
  - they express sounds and patterns of sound that by the age of three or four is recognisably music with rhythm, repetition, melody and simple harmony;
  - they make marks and images that are expressive of what they see and how they feel about it, both imitative and with growing control of aesthetic form;
  - they move and explore what they can do with their bodies, including those aesthetic elements of shape, rhythm and repetition … in other words dance;
  - they create stories and make narratives of the world as they see it, or wish it to be, and other worlds and people that might (hypothetically) be; and
  - they step into the shoes of other people, pretending to be both the people they see around them, and other people and worlds that might (hypothetically) be, to see how they might work … in dramatic play they use their bodies and language to make *models* of how people and the world behave.

This iteration might be getting tedious, as it emphasises just how close and not merely complementary but integrally interwoven the basic principles of science and arts are to children … and what is often overlooked and underrated is the high level of untutored sophistication of their experiments in both.

The key italicised word that readers might have noticed in both paragraphs is *models*. Models are of course ubiquitous to learning in science, whether three-dimensional, diagrammatic, algebraic or now more commonly virtual, and they are all designed entirely for cognition and cognitive understanding. That is fine for the science of the external world. But how about for understanding human behaviour?

That is where we come to drama, the art form and pedagogy of this book. Models exploring and mapping human behaviour must take into account the emotional, the sensory, the kinaesthetic and the social elements and components of that behaviour, not instead of, but as well as the cognitive. The models that drama automatically creates do exactly that. In stepping into a drama, whether as actor in a play, participant of role-play or dramatic play, or as audience, we step into a fictional model world that can be made as like our real world as we choose, three-dimensional, fully embodied and as real-world authentic as we want it to be. In this fictional world, through drama's unique quality of the voluntary suspension of disbelief (Coleridge, 1817)—in other words by agreeing to pretend and enter that world together—we can experience the physical actions, the thoughts, ideas and the real emotions of its inhabitants (the 'characters') ... with two necessary differences from the real world:

- because the dramatic world is in our hands (or those of the playwright and actors), and like any good and useful learning model it can be manipulated, altered, modified or changed to see 'what would happen if...' as well as 'what's happening?', 'what comes next?', 'why?' and 'what's at stake here?'
- because the world is pretend, we can step out of the fiction, without any real-world consequences—we can replace empathy with emotional distance and leave the strong emotions, the deep relationships, the fixed attitudes and behaviours of the characters in our model world completely behind us as memories, and more important, as considerations for reflection.

That is the learning in drama. It is a constructivist and reflective learning, not just filling an empty jar with 'new' content but scaffolding on what the students already understand through experience, through their bodies, minds and emotions; as the educational drama pioneer Dorothy Heathcote, who is much quoted in this book, famously observed: 'to literally bring out what children already know but don't yet know they know' (reported in Wagner, 1975, p. 13).

With these integral ties from birth onwards, to continue in schools to teach science through drama would seem to be a no-brainer, but our Western cultural heritage, and its traditional dichotomy into 'the two cultures' (Snow, 1959), has determined the shape of the school curriculum and timetable, and put the two in opposition, further siloed away from each other both by subject and learning area distinctions and by hierarchical considerations that raise the status of one and depress the other. So, any examples of science and drama taught together, instead of being normal practice, are as scarce as hen's teeth, more so as students get older. The editors of the book explain why in detail. Partly because it is so rare, breaking down the silos demands imagination and collaborative courage. The editors also explain with an elegant scientific metaphor why for the really effective teaching of drama and science together, knowledge and understanding both of the science and the art form of drama that is being used as the pedagogy, are equally necessary.

The diversity of the book offers an interesting extra challenge. Its origin is the reason why it is so diverse: an open invitation, sent internationally, to contribute. Most of the contributors share the editors' frames of reference, but not quite all. The 'healthy symbiosis' of science and drama teaching expertise that the editors encourage, and that they demonstrate in their own contributions, is a rare phenomenon. This is at least partly because of the imbalance of training and understanding of the two disciplines within our education systems. All the contributors without exception share the editors' vision and mission, to provide engaging, embodied and experiential learning, that will help students learn holistically not only new knowledge, but more importantly to understand 'what they know, but do not know they know'. However, several contributors have found striking alternative routes to their students' holistic engagement: through instinctively recognising the dramatic possibilities of puppets; through mimetic exercises; through psycho-social exercises; through training role-plays; and one through internet game-theory, from which the authors have constructed an almost classically simple example of educational drama in an outdoor setting. These alternative 'left-field' approaches give a valuable extra dimension for the reader, and they will provide very useful comparison with the more orthodox symbiosis of expertise demonstrated by the majority of the exemplars … very useful for the critical readers to examine against that practice and their own. The contributions demonstrate the necessity of collaborative courage, evident where the contributors have taken that crucial step to bring their own discipline and expertise together with one from the 'other culture', to find a dynamic complementarity.

The most important aspect of this valuable book is its very existence, which can only raise the desperately needed awareness of schools and teachers of the congeniality and kinship of sciences and the arts. It provides thought-provoking materials for any imaginative, collaborative and open-minded teacher to say … 'Yes of course … we really should … there's a great idea we could have a go at … our students REALLY need this!'

Brisbane, Australia
February 2021

Professor John O'Toole AM

John O'Toole was Foundation Chair of Arts Education at the University of Melbourne. He has been teaching and working in drama, theatre, and education for half a century, with all ages and on all continents. In 2014, he received an Order of Australia Award (AM) for services to drama education.

References

Coleridge, S. T. (1817). *Biographia literaria.* Chapter XIV. Variously published.

Snow, C .P. (1959). *The two cultures.* Rede Lecture, Cambridge University.

Wagner, B.-J. (1975). *Dorothy Heathcote: Drama as a learning medium.* Portsmouth NH, Heinemann.

Acknowledgements

The editors thank all chapter authors for their contribution and energy in bringing this book into being.

We enacted a process of peer review and created a community of practice, showing care and consideration through some trying times during the global pandemic of 2020.

Gratitude to our colleagues Professor John O'Toole AM, Professor Russell Tytler, and Professor Vaughan Prain for providing guidance and feedback. Appreciation to Marie Christodulaki, our proof-reader extraordinaire.

Our community continues, and we welcome you to join us. Please explore the additional materials provided by the authors at: https://blogs.deakin.edu.au/science-drama/.

Contents

Contributors

Dayle M. Anderson Victoria University of Wellington, Wellington, New Zealand

Delia Baskerville Victoria University of Wellington, Wellington, New Zealand

Jane Bird Melbourne Graduate School of Education, University of Melbourne, Melbourne, Australia

Leni Brown University of South Australia, Adelaide, South Australia, Australia

Tricia Clark-Fookes Creative Industries, Education and Social Justice Faculty, Queensland University of Technology, Brisbane, Australia

James P. Davis Faculty of Creative Industries, Education and Social Justice, School of Teacher Education and Leadership, Queensland University of Technology, Kelvin Grove Campus, Brisbane, Australia

Susan Davis Central Queensland University, Rockhampton, Australia

Joseph Ferguson Faculty of Arts & Education, School of Education, Deakin University, Burwood, Australia

Robyne Garrett University of South Australia, Adelaide, South Australia, Australia

Shelley Hannigan Faculty of Arts & Education, School of Education, Deakin University, Waurn Ponds, Australia

Sari Harmoinen Faculty of Education, University of Oulu, Oulu, Finland

Senka Henderson Creative Industries, Education and Social Justice Faculty, Queensland University of Technology, Brisbane, Australia

Donna King School of Education, Faculty of Education & Arts, Australian Catholic University, Brisbane, Australia

Debra McGregor Oxford Brookes University, Oxford, UK

Debbie Myers University of Northumbria, Newcastle, UK

Lisa O'Keeffe University of South Australia, Adelaide, South Australia, Australia

Kathryn Paige University of South Australia, Adelaide, South Australia, Australia

Vaughan Prain Faculty of Arts & Education, Deakin University, Waurn Ponds, Australia

Jo Raphael School of Education, Deakin University, Burwood, VIC, Australia

Richard Johnson Sallis Melbourne Graduate School of Education, University of Melbourne, Melbourne, Australia

Pirkko Siklander Faculty of Education, University of Oulu, Oulu, Finland

Carolyn Julie Swanson School of Education, Auckland University of Technology, Auckland, New Zealand

Russell Tytler School of Education, Deakin University, Burwood, VIC, Australia

Kitty van Cuylenburg School of Education, Deakin University, Burwood, VIC, Australia

Peta J White School of Education, Deakin University, Burwood, VIC, Australia

Sparking Learning in Science and Drama: Setting the Scene

Jo Raphael, Peta J White, and Kitty van Cuylenburg

Abstract In this chapter we 'set the scene' for the diverse science-drama links represented in this book. We consider the various approaches to science and drama and how they are combined, aligned, infused and adapted to generate understandings, insight and engagement in learners. We begin by looking broadly at the discipline areas of arts and sciences, focusing on what they share. Drawing from our personal experiences, we test and reject the notion of a dichotomy. We argue that bringing drama together with science helps students think across the boundaries of disciplines. We recognise that when planning for the future of education it is necessary to understand that students of today live in a time of unprecedented social, economic and environmental change and challenges, alongside opportunities presented by an explosion in scientific knowledge. We explain the ways that bringing science and drama together can help to develop the skills, attitudes, values and creativity that young people will need to thrive in and shape their world. In this chapter we acknowledge that pedagogy of science drama is under theorised and explain how each chapter within this volume makes a contribution towards understanding and exemplifying science and drama as pedagogy. We identify various science/drama models, and propose symbiosis–mutualism, commensalism and parasitism—as a way of understanding the relationship between drama and science disciplines when connected for learning. In particular, we unpack the idea of symbiosis, with a focus on what each discipline can bring to the other. Ultimately, we describe our temporal pedagogical model of intertwined synergy for sparking and generating learning in science and drama.

Keywords Science · Drama · Interdisciplinarity · Transdisciplinarity · Symbiosis · Pedagogical models

J. Raphael (✉) · P. J White · K. van Cuylenburg
School of Education, Deakin University, Burwood, VIC, Australia
e-mail: jo.raphael@deakin.edu.au

P. J White
e-mail: peta.white@deakin.edu.au

K. van Cuylenburg
e-mail: k.vancuylenburg@deakin.edu.au

P. J White et al. (eds.), *Science and Drama: Contemporary and Creative Approaches to Teaching and Learning*, https://doi.org/10.1007/978-3-030-84401-1_1

Introduction

Imagine if there was a learning spectrometer that was capable of detecting flashes of learning and moments of intense illumination in the mind. We imagine this learning spectrometer would be detecting these flash point moments being sparked in learners when the disciplines of science and drama are brought together. As educators and editors of this book, we share a passion for learning and a quest for understanding the kinds of pedagogical approaches that can lead to the most inclusive and powerful learning potential for all learners. The chapters in this book are all, in one way or another, engaging in this quest for understanding what happens in learning and teaching when drama and science are brought together. In this chapter we 'set the scene' for the examples in this book, considering the various approaches to science and drama and how they are combined, aligned, infused and adapted to generate understandings, insight and engagement in learners. We identify various science/drama models, and ultimately, we describe our temporal model of intertwined synergy for learning in science and drama.

Background

To provide some context for this volume of contemporary approaches to teaching and learning in science and drama, we take a moment to examine the origins of our interests in both science and the arts as well as what brought us together as editors on this project. As educators with strong backgrounds in our discipline areas, we are considered to hold expert knowledge in one area but find ourselves reluctant to be confined to one box. Hildegard of Bingen lived around a thousand years ago, and is considered a polymath. She is celebrated for her insights and discoveries in biology, medicine and natural history and equally for her music composition and play-writing that explored and presented her understandings of the world (Mark, 2019). Like Hildegard, we love both sciences and arts, we see the benefit of bringing them together, particularly the dramatic arts. Hildegard did not feel the need to confine her studies to one area, and we agree that in order to attain expert status in one discipline area, you need not let go of all others. We want to join the boxes together, jump between them, dwell in them and play amongst them, to see what might be possible.

As editors, we have each reflected back on our school days and the pressure to narrow our focus and specialise as we progressed through to the senior years. We remember the stress involved in deciding which subjects to take and which to let go, because of our curiosity and interest in learning across multiple areas. For each of us the direction we finally took was determined to some extent by chance and yet we held onto our other curiosities and passions in different ways. Fortuitously, our shared interests led us to the circumstances that brought us all together and generated the idea for this edited collection.

At the end of secondary school, Jo found herself narrowly choosing studies in drama and education over health sciences for her first degree. While a career in health sciences seemed the 'sensible' choice, her love of drama and performing led her to audition for a drama course noted for being in high demand, with few available places. She was offered and accepted a place that seemed too good to give up, and without compromising her theatre major and complementary literature minor in the degree, she continued to squeeze in an additional minor in environmental science. She has since taught drama and theatre in schools and universities for over 30 years, while still facilitating and making theatre as artistic director of Fusion Theatre (www.fusiontheatre.com.au). Applied theatre work and drama for inclusive education became the focus of her Ph.D., and she continues to explore drama and theatre as pedagogy applied to learning across all disciplines.

Peta had similarly broad interests and capabilities as a secondary school student, including science, the arts (music, photography and drama) Then one day in year nine she was asked to produce an insect collection. She made a call to her Mum asking her to send some insect bodies and Dad's old shirt box. Her Mum lived on a farm with a large front window and no curtain and many insects died each night, drawn in by the light. The insect collection received 11/10, Peta won the science prize for that year, and an identity that now fore fronted 'science'. Peta was successful in receiving entry to a prestigious photography course but instead chose a Bachelor of Science with a major in zoology. Her interests in the arts remained and diversified but came into sharp focus in 2014 when working with Jo in a research group applying Self-study in Teacher Education methodology (Hannigan et al., 2016; Raphael et al., 2016; White et al., 2020). Interesting fun fact–both Kitty and Jo are synesthetes, which makes working with them… colourful!

Kitty has worked professionally in mine-site hydrogeology and as a performer and is currently theatre-maker and board member of In The Park Productions (https://www.intheparkproductions.com.au/). She undertook a Master of Teaching with Teach for Australia in 2016, has since taught across both science and drama, and currently works as a teaching and learning advisor with Teach For Australia. Teaching and learning across the scientific and artistic is who she is–as a little girl she would write poetry about the water cycle, in adolescence she would create movies about forensic science, and as a young adult she produced theatre about mental illness. Kitty majored in geology and drama as part of her undergraduate Bachelor of Arts/Bachelor of Science (Honours) and has since pursued every opportunity to combine the two, often in science communication events (Laneway Learning, Pint of Science and TedX).

While studying the Master of Teaching, Kitty chose the research pathway and undertook a project in interdisciplinary practice of science and drama under the co-supervision of Peta and Jo (Chapter 'The Science Drama Project: Meaning in the Middle'). It was during this project in 2016 that we three agreed there was a need for a book to bring together research in science and drama pedagogy with examples of exemplary practice. When we presented the research and talked about our practice, teachers and academics asked us, 'Where is the book on science and drama?'. This book of researched examples, along with the corresponding website of complementary resources (https://blogs.deakin.edu.au/science-drama/) is our response to this perceived need. The book includes examples of science-drama links of relevance to academics, teacher educators, pre-service teachers and teachers in schools. We cast the net widely, putting out the call for contributions to educators in both the drama and science communities, and we received a diverse range of chapters from contributors with wide ranging experiences. We hope that you all benefit from the community of practice we are collectively building as we move towards better understanding the work of bringing science and drama together for learning.

Science and Arts—Branches of the Same Tree

Science and the arts share much in common, they are both ways that humans experience and make sense of the world. Science, like the arts is "inspired by emotion and passion, shaped by intuition, and finds expression in narrative and aesthetic forms" (Nicholson, 2011, p. 196). They both seek elegance and are both charged by wonder. Consider a photographic image of the planet Earth from space, there is a beauty perceived by both artists and scientists, and it is this sense of beauty that motivates learning (Nicholson, 2011; Winston, 2010). Iconic images such as this can "provoke an openness and a receptivity to our deepest emotional capacities and a connection between those and the dynamics of the broader social-ecological systems in which we operate" (Westley & Folke, 2018, p. 30). Aesthetic appreciation is an affect, potentially prompting wonderings about life on Earth, impacts of humans on the planet, and issues of global environmental sustainability. In the pursuit of new knowledge and understandings, both science and drama require imagination, they play with ideas and employ experimentation in the quest for new discoveries. Both engage with aesthetic knowing, rather than anaesthetic, making use of all senses, rather than dulling them. Humans are living, sensing, feeling beings, we are 'live creatures' as Dewey (1934) reflects in his book Art as Experience, suggesting the difference between art and science is simply where the emphasis falls:

> … in the constant rhythm that marks interaction of the live creature with his surroundings. … The odd notion that an artist does not think and a scientific inquirer does nothing else is the result of converting a difference of tempo and emphasis into a difference in kind. (p. 15)

With education in mind, Dewey explains that scientific and artistic modes of inquiry, although having different tempos, are both part of the rhythm of learning, and both are necessary for the way they deepen our experience and understanding of the world around us. This is a "a kind of 'wholeness engagement' where emotions, intellect, and practice react together simultaneously" (Abed, 2016, p. 170). Wholeness engagement could also describe the nature of Indigenous science whereby First Nations peoples have built empirical knowledge of the natural world through long-standing close connection to local areas, or 'country' (Australian Research Council,

2008). Indigenous Scientific Knowledge is increasingly important in contemporary times. For example, detailed knowledge of controlled burning tested over time and passed through generations by Aboriginal people, is recognised as a way of reducing risk of devastating bush fire. It is easy to imagine this Indigenous Scientific Knowledge coming from Dewey's 'constant rhythm' of learning in connection with surroundings, through aesthetic and embodied practice and social learning, with a sense of responsibility to the sustainability of nature, and which is often communicated through the arts.

Commonalities in arts and science are also noted by Jung, who defined the 'artist-scientist' as one of several human archetypes (Westley & Folke, 2018). The artist-scientist is driven by wonder and curiosity, is a creator and a discoverer, is intuitive, imaginative and an improviser. Creativity and innovation are traits valued in both artists and scientists who require the "courage to imagine the unimaginable" (Turkka et al., 2017, p. 1). We have already mentioned Hildegard of Bingen, but there are many other well-known historical examples of artist-scientists whose achievements have been made through bringing together these ways of perceiving the world. Leonardo da Vinci was an artist and a scientist, an inventor and a philosopher, and each facet enhanced his ability to go deeper into the other (Capra, 2014). Albert Einstein is less known for his considerable artistic talents than his achievements as a theoretical physicist, but valued his artistic abilities highly and believed they contributed to his scientific discoveries. Einstein saw the power of imagination as crucial and arts and sciences as branches of the same tree (Caglioti, 2017).

There has been a tendency for the arts/humanities and sciences to be seen as 'two cultures', and this division is unhelpful for solving the problems of the world (Snow, 1959). The separation of discipline areas is intensified by school curricula with clearly defined key learning areas, their own content descriptions and achievement standards. Key learning areas like the arts and the sciences are further divided into subjects. These separate curricula are then developed by people who are usually considered experts in and advocates for that particular discipline. Secondary school teachers are registered in two curriculum areas, often with one major area in which they focus their practice. In most secondary schools it is uncommon for learning areas to co-join. The disciplines are sometimes in competition for time and resources particularly within what is often referred to as the crowded curriculum.

Segregating discipline areas may provide a sense of clarity and that could be interpreted as security, for some. However, these 'silos' (that are so pervasive in education), can provide a barrier for the kinds of learning that promote deeper understanding amongst diverse learners (Byrne & Mullally, 2016), as discussed earlier in this chapter, a siloed approach does not reflect the complexity of the real world and the spectrum of thinking required to solve problems in our rapidly changing environment–particularly for 'multipotentialite' students who have diverse interests and multiple possible careers across their lifetime (Wapnick, 2015).

There has been a movement to add the arts into STEM (Science, Technology, Engineering and Mathematics) creating the acronym STEAM, the name itself suggestive of a driving and energetic force. A STEAM approach acknowledges that finding solutions to complex problems requires thinking across disciplines, and that the arts can spark creativity and imagination for problem-solving and innovation (OECD, 2019). A meta-analysis of STEAM programs revealed that students experience positive outcomes from involvement in STEAM, including confidence in problem-solving, identity as scientists, entrepreneurship and appreciation of teamwork (Kang, 2019). However, the same study showed that research is needed to inform more meaningful interdisciplinary activities, and in preparing teachers for effective collaboration across disciplines (Kang, 2019).

The examples of science and drama in this book grapple with curriculum and the question of which discipline in the curriculum their lessons serve. Sometimes it is the science curriculum, sometimes the drama curriculum, and in some cases both. Often in these examples, the learning moves beyond the disciplines of science or drama to other parts of the curriculum, to what are known in the Australian Curriculum as the General Capabilities, including Personal and Social Capability, Critical and Creative Thinking, and Ethical Understanding (ACARA, 2021). We are interested in the ways that interdisciplinarity (science and drama combined) can spark learning that could not be typically addressed through either disciplinary practice. We believe that this transdisciplinary practice benefits learners through access to affective learning (Chapter 'Transdisciplinarity: Science and Drama Education Developing Teachers for the Future'), or positive emotion (Chapter '"This is the Funniest Lesson": The production of Positive Emotions During Role-Play in the Middle Years Science Classroom'), or enjoyment and laughter (Chapter 'Dramatic and Undramatic Emotional Energy: Creating aesthetic and Emotive Learning Experiences in Science Classrooms'), or aesthetics learning that often occurs through practical engagement (Chapters 'Dramatising the S and the M in STEM', 'Ice Age is Approaching: Triggering University Students' Interest and Engagement in Gamified Outdoor Playful Learning Activities' and 'Science, Drama and the Aesthetic').

Our arrangement of these chapters is recognition of this curriculum challenge—due to the intertwined nature of the two we felt it would be misleading to arrange them by discipline or pedagogy. Instead, we identified triads or couplets of research which, in order, move from the panoramic to the macro 'photographic zoom' of how these two disciplines may present. We have structured for perspective rather than curriculum. Through this chapter we lean into dramatic pedagogies and how they gracefully and mutually intertwine with science for learning. We establish the historic and theoretical panoramic (Chapters 'Mutuality and Inter-relativity of Drama and Science', 'The Science Drama Project: Meaning in the Middle', 'The Treatment of Dr. Lister: Investigating and Revisiting a Famous Primary Science-Based Drama'), take in the landscape of practice in the classroom (Chapters 'Dramatising the S and the M in STEM', 'Responding to Climate Change: Developing Primary Children's Capability to Engage with Science Through Drama', 'New Eden: Mediating Pre-service Teachers' Conceptions of Education for Sustainable Development'), the portraiture of ethical and environmental deliberation (Chapters 'Transdisciplinarity: Science and Drama Education Developing Teachers for the Future' and 'Art-Science Education in the Anthropocene: Embodied Metaphor with Puppets and Performance'), macro photography of inner worlds (Chapters '"This is the Funniest Lesson": The production of Positive Emotions During Role-Play in the Middle Years Science

Classroom', 'Dramatic and Undramatic Emotional Energy: Creating Aesthetic and Emotive Learning Experiences in Science Classrooms', 'Does Being Positioned in an Expert Scientist Role Enhance 11–13 Year-Old Students' Perceptions of Themselves as Scientists?') and a final gallery walk of stories (Chapters 'Stories from History: More Authentic Ways of Thinking Through Acting and Talking About Science' and 'Australian Women in Science: A Model for a Research-Based Theatre Project in Secondary School Classrooms'). Chapter 'Science, Drama and the Aesthetic' book-ends the work by leaning into the scientific but completes this journey by inviting us to see these chapters anew, through our own eyes with an aesthetic appreciation across both disciplines.

Learning About Science in and Through Drama

In most of the chapters in this book, drama is employed as a pedagogical approach to engage students, to enhance learning in science and students' attitudes towards science as a worthwhile endeavour. In most cases learning in drama is subservient to the science, implying a lack of reciprocity in the relationship. While we are in favour of using drama as pedagogy in an integrated approach to learning, we are aware of criticism in the arts and drama education community of the use of drama simply in service of other learning areas, especially if its use does not contribute to the study of drama as a discipline (for example Hornbrook, 1989). Dinham (2019) describes this as a 'handmaiden' approach when the arts, and in this case drama, is used to support other areas with little or no attention to learning in the discipline of drama. Dunn and Stinson (2011) critique the lack of aesthetic understanding in drama pedagogy when teachers, who may lack knowledge and experience in drama, use it in an unsophisticated way, such as simplistic role-plays, without fully appreciating the potential of the art form. Successful integration involves careful consideration of the elements and aesthetic outcomes for drama as well as other subject outcomes (O'Toole, 2017).

We believe drama is a powerful pedagogy, and we are committed to the belief that more teachers should be encouraged to draw upon it as a way of activating learning. While drama is a way of learning and understanding the world, like science, drama is a rich body of knowledge with its own skills, techniques and processes deserving of its place in the school curriculum. Students need to have opportunities for learning 'in' as well as 'through' drama.

Drama and Science Together for 21st Century Education

The Organisation for Economic Co-operation and Development (OECD) launched The Future of Education and Skills 2030 project (E2030) with the aim to help countries find a shared vision in response to two broad questions:

- What knowledge, skills, attitudes and values will today's students need to thrive and shape their world?
- How can instructional systems develop these knowledge, skills, attitudes and values effectively? (OECD, 2018, p. 2)

The E2030 project acknowledges that we are educating in a time of unprecedented social, economic and environmental challenges. We are experiencing an explosion of scientific knowledge, "creating new opportunities and solutions that can enrich our lives, while at the same time fuelling disruptive waves of change in every sector .... It is time to create new economic, social and institutional models that pursue better lives for all" (OECD, 2018, p. 3). With an eye on the future of education, the OECD suggests that education can make the difference, but will require radical changes to curricula. We suggest this must also apply to the way we teach. Learners, E2030 project claims, will need to mobilise a broad set of "knowledge, skills, attitudes and values to meet complex demands" and the capacity to "think across the boundaries of disciplines and 'connect the dots'" (OECD, 2018, p. 5).

Many of the examples in the chapters in this book offer innovative and creative approaches to education, working to interconnect across disciplinary boundaries. They frequently draw upon multiple modalities, sometimes providing content and structure and sometimes providing more creative freedom, because this is one of the things an arts-based approach can do: "allow the crazy dream modality its freedom and inventiveness. ... 'Let us learn to dream!'" (Abbs, 1989, p. 23).

Problem of Under-Theorised Pedagogy of Science and Drama

Educators who have brought drama and science together for learning have been keen to describe the practice in order to share it with others. However, Braund et al. (2013) suggest more research and theorisation is needed in order to understand how drama works for science learning and to assist teachers to use it to its full potential (p. 6). In a study of preservice teachers' use of drama, they noticed the challenges that could occur when the teachers failed to make adequate links between drama pedagogical content knowledge and science knowledge (Braund et al., 2013). They recognised that science teachers often lacked drama pedagogical expertise (and sometimes also lacked science knowledge), and teachers with drama pedagogical knowledge often missed opportunities to link science concepts to drama activity. They also observed drama activity being overly teacher directed in a way that was controlling and discouraging of learner autonomy. Furthermore, there was a lack of critical evaluation and reflection in the drama as discussions were closed down and students were given little opportunity to negotiate meaning in shared tasks. Braund et al. (2013) concluded that for "drama to be successful, and reach its full potential, some key pedagogical principles have to be established ... these include better linking of the features of drama to phenomena, concepts and processes and providing a higher degree of learner autonomy in devising, presenting and critically evaluating what is presented" (Braund et al., 2013, p. 12).

We agree that greater theoretical understanding of drama for science learning is required in order to maximise the opportunity for learning. The authors contributing to this book have all researched science drama practice that they believe has value and is worth sharing. Many have also identified challenges and practices best avoided. All chapters are working towards describing, analysing and presenting a pedagogy of science and drama to inform other teachers. We have read the chapters with the view to understanding the diverse ways the authors bring science and drama together, and in this chapter we tease these out to see what pedagogical models are implied by the approaches, and what lessons these might offer for a productive synthesis.

Towards Pedagogical Models for Science and Drama

Nicholson (2011) reminds us of the challenge of maintaining "an even balance of power between arts and sciences, in which the educational imperative extends, rather than diminishes, their capacity to move, excite, and provide an imaginative vision of the future" (p. 197). In this section, we are exploring the nature of this balance and how points of balance may serve educational imperatives through looking at various models. Bringing together two discipline areas such as science and drama inspires the application of spatial models to describe the relationship as either multi-disciplinary, interdisciplinary, or transdisciplinary. Representing each discipline as discrete is the established (traditional) way the school curriculum delivers subjects individually (Fig. 1). In multi-disciplinary models each discipline is located in proximity, often as a contrast, and more of an addition rather than an integration. Interdisciplinary learning can be represented by a Venn diagram (Fig. 1) where disciplines remain separate except for the middle/overlapping area which is highlighted as the 'third colour' (Akkerman & Bakker, 2012).

The generation of the third colour represents a new opportunity for learning as each discipline connects and can be seen in the way schools may implement STEM curriculum (especially those also combined with the arts into STEAM). The interconnection of each discipline aims to trigger learning through 'boundary crossing' (Akkerman & Bakker, 2011) which may otherwise remain inaccessible in specific disciplinary work. Interdisciplinary learning is the model for some of the chapters presented herein, with the tools of drama crossing the boundary into science and the scientific inspiring the imagination in drama to create a combined outcome. Myers (Chapter 'New Eden: Mediating Pre-service Teachers' Conceptions of Education for Sustainable Development') and Baskerville and Anderson (Chapter 'Responding to Climate Change: Developing Primary Children's Capability to Engage with Science Through Drama') both present researched examples where dramatic conventions hold space and cross boundaries with science.

A transdisciplinary approach is defined as "an approach to curriculum integration which dissolves the boundaries between the conventional disciplines and organises

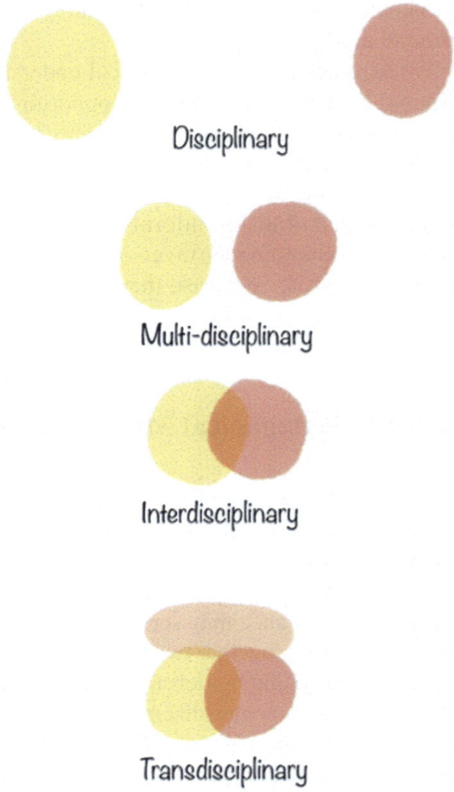

Fig. 1 Spatial models representing fields as disciplinary, multi-disciplinary, interdisciplinary and transdisciplinary, generating a new space as drama and science concepts and practice cross boundaries generating new learning

teaching and learning around the construction of meaning in the context of real-world problems or themes" (UNESCO, n.d.). Transdisciplinary models (Fig. 1) create a new space which focuses on learning or through-lines across both disciplines, owned by neither, resulting in wholistic integration. Hannigan and Ferguson (Chapter 'Art-Science Education in the Anthropocene: Embodied Metaphor with Puppets and Performance') present an example through the use of embodied metaphor across both arts and science with use of objects as students create and bring animal puppets to life for 'theatre in a suitcase'. Van Cuylenburg (Chapter 'The Science Drama Project: Meaning in the Middle') demonstrates transdisciplinarity as drama and science meet in the middle, creating a space both pedagogical and physical, where students can explore ideas from their perspectives and relevant to their lives. Raphael and White (Chapter 'Transdisciplinarity: Science and Drama Education Developing Teachers for the Future') also explain how controversial science issues could be explored by applying drama pedagogy accessing emotional and affective learning towards a more comprehensive understanding of complex ethical issues not likely achievable without a transdisciplinary approach. In fact, much of the research presented in this book focuses on interdisciplinary and transdisciplinary approaches and we argue that the intertwining of both disciplines (Menken & Keestra, 2016) enables sparks of learning science and drama and often meaningful learning in areas that transcend these two disciplines. It is almost like science and drama undergo a chemical change when properly used, into a fusion of constructivist learning.

We want to further unpack the transdisciplinarity of drama and science teaching and learning in combination, through applying a temporal model. Tytler et al. (2019) applied temporality to model the epistemic practices for interdisciplinary STEM paying attention to the cultural and historical nature of each discipline. They also apply consideration of timescale in the model where micro (moment by moment application of each discipline), meso (planning discipline specific learning over series of lessons), or macro (considering the learning progression across a longer period such as a year) levels interact resulting in richer learning through explicit planning at these different time scales. Clarke-Fookes and Henderson (Chapter 'Mutuality and Inter-relativity of Drama and Science') refers to the work of Lerhman (2014) to present a model that also explores interdisciplinarity over time using the term 'hiking the horizontal'. They reject the polarities, binaries or hierarchies of combining disciplines and illustrate the paths by which two disciplines may co-exist. Like a sonic frequency, the practitioner may dive into one discipline, surface into the other or swim the boundary of both. Lerman suggests that "In a hierarchical order, the only way to make a distinction is to literally put the other idea 'down'. There is only room for one idea. But in hiking the horizontal, many ideas can coexist" (Lerman, 2014, p. xvi). We found this idea (hiking the horizontal) to be a revelation and modelled it to explore how drama and science might be taken up across the micro, meso, and macro levels during interdisciplinary or transdisciplinary learning (Fig. 2).

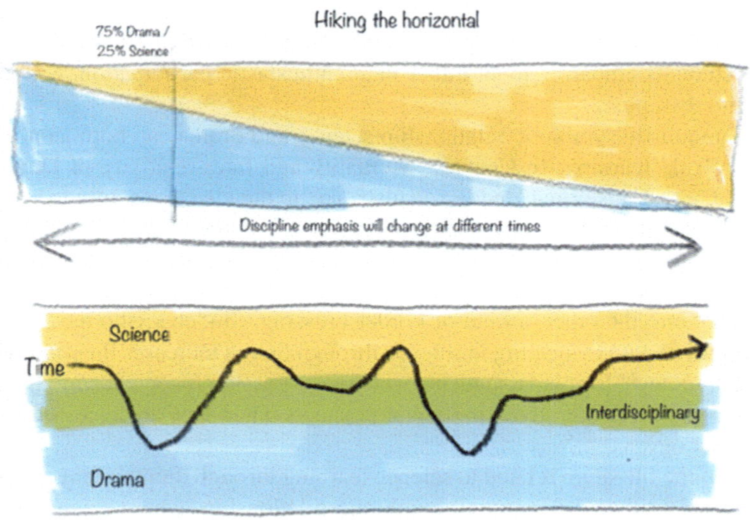

Fig. 2 Models developed to explore 'hiking the horizontal'—a temporal consideration in interdisciplinary and transdisciplinary teaching and learning

Drama and Science: A Symbiotic Relationship

In thinking about teaching and learning in drama and science we often applied metaphors. Perhaps not surprisingly, these metaphors sometimes came from science, such as the idea of drama and science in education as a form of symbiosis, the close relationship between two species in which at least one benefits (Merriam-Webster, 2021). Take for example, the clownfish and the sea anemone, who share a mutual symbiotic relationship. The clownfish feeds on small invertebrates that otherwise harm the sea anemone and the sea anemone gains nutrients from the faecal matter (digested invertebrates) of the clownfish. The clownfish is protected from predators by the anemone's stinging cells (to which the clownfish has become immune as they develop a mucus layer to protect them after the first sting), and the sea anemone is protected by the high-pitched sound emitted from the clownfish that scares off butterfly fish that might otherwise eat it and is cleaned as the clownfish eats the dead anemone tentacles. In short, clownfish and sea anemones nourish and protect each other.

By expanding the notion of symbiosis, we can find nuances to our understanding of the symbiotic relationship through which drama and science relate. Several different kinds of symbiotic relationships are defined, and we have chosen three to apply in our metaphor: mutualism, commensalism and parasitism. In the mutualistic symbiotic relationship, both science and drama benefit from their relationship with the other. In commensalism, one will benefit while the other is unchanged. In parasitism, one benefits while the other is harmed. We represented these in Fig. 3. We certainly would not choose parasitic symbiosis in science and drama teaching and learning, although we acknowledge it may happen. For example, drama that purports to be about science, but is poorly researched and gets the science wrong, would be doing harm to students needing to learn science. Similarly, science that uses drama in careless ways, perhaps causing students embarrassment rather than enthusiastic participation, could bring drama into disrepute.

In a mutualistic symbiotic relationship science and drama can both gain as partners in holistic learning. In Chapter 'Mutuality and Inter-relativity of Drama and Science', Clark-Fookes and Henderson also take up the idea of a symbiotic relationship. They explain that science, with its "emphasis on rationality, empirical inquiry and discursive expression is aided by the affective understandings and non-discursive representation of ideas in drama, and vice versa". In addition, one can be a gateway for students into the other, a kind of border-crossing. Students who have a love of drama might find that engaging in science through drama increases their interest and learning in science. Similarly, students who have a passion for science might engage in and learn more deeply about particular concepts when they are accessed through drama pedagogy.

Much of the literature related to science learning through drama seems to suggest a commensal (rather than mutual) symbiotic relationship, where science benefits and drama is unchanged. Working towards a theoretical model for science drama, Abrahams and Braund (2012) suggest a 'drama space' to move students from the

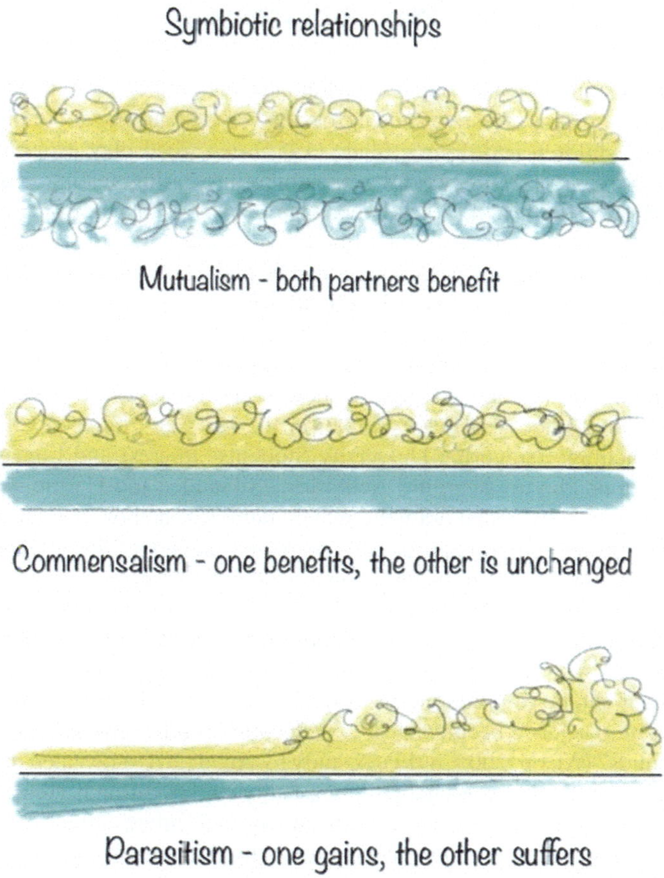

Fig. 3 Models of symbiosis used to explore the relationship of science and drama working together for learning over time

'learners world of knowledge' to the 'science world of knowledge'. Braund et al., (2013, p. 6) develop the idea of drama space as a space of 'cognitive dissonance' which is "the 'distance' between the two worlds of knowing, and the 'experiential space' (in this case filled by drama)" and drama activity can be employed by the teacher "to reduce dissonance and close the gap between the two worlds" (Braund et al., 2013, p. 6). Understanding the types of drama that can be drawn upon, and the purpose and potential of these approaches, will assist teachers when designing science drama programs. Beyond bridging the gap, or facilitating knowledge transfer, there are numerous ways in which science benefits from drama which we consider in the following sections.

Drama, Engagement and Playful Learning

Young children spontaneously engage in dramatic play either individually or with others, as their own very effective and engrossing way of learning about their world. In his study of children and drama, Slade (1954) described play as a child's way of "thinking, proving, relaxing, working, remembering, daring, testing, creating and absorbing" (Slade, 1954, p. 42). The advantage of play is that it allows for experimentation with ideas 'as if' real, but without real world consequences. Therefore, through drama, daring and dangerous ideas can be played out in the safe space of the fictional frame of the drama. Play is imagination in action, according to Vygotsky (1976), and can give rise to big ideas. The ability to test big ideas is important for scientific discovery and innovation. Through drama we can harness the power of play, for students of any age, to explore themes and ideas as we plan to achieve particular learning outcomes. For example, in the Ice Age drama in Chapter 'Ice Age is Approaching: Triggering University Students' Interest and Engagement in Gamified Outdoor Playful Learning Activities', Siklander and Harmoinen engage university students in extended dramatic play to demonstrate how deep learning can be 'triggered' through playful approaches, no matter the age of the students.

The potential for drama to deeply engage students in learning about science is often mentioned in the research literature. Factors that contribute to this engagement include the diversion from what some would deem more traditional approaches to learning science, with a focus on transmission of knowledge. For some learners, science content can be intimidating in its complexity and "the fragile formation of an identity within science becomes lost … Education works best when it combines hearts and minds" (Alsop, 2005, p. 17). A study by Abed (2016) revealed that "drama is a powerful strategy that can change 'completely' the traditional view of science class from being rigid and boring to being lively and entertaining" (p. 170). We see playful modes of learning in most chapters in this volume. Henderson and King (Chapter '"This is the Funniest Lesson": The production of Positive Emotions During Role-Play in the Middle Years Science Classroom') describe research that reveals how positive emotions in the science classroom enable student engagement and enjoyment even in the study of serious socio-scientific ethical dilemmas associated with organ harvesting and transplants. Far from being trivial, a sense of fun and joy that dramatic play brings to science could be an important factor that makes the study of science seem less intimidating, holding the interest of students who might otherwise see science as not being for them.

Drama for Embodying Science Learning

Engagement is often attributed to the embodied nature of learning in drama that moves students away from chairs, desks and tables. Students are sometimes taught as if what goes on in their heads is the only thing that matters. The Cartesian notion of mind–body dualism has long been questioned and replaced by the understanding that we experience the world and come to know it through mind and body, and yet (Alsop, 2005) suggests the tendency to favour reason above emotion persists in science. The effect of 'affect', what is felt before it is thought, is gaining increasing attention for its importance in learning. Emotions, according to Alsop (2005), "have a considerable influence over what happens in the classroom. Some (such as joy, love, happiness and hope) act to enhance education, optimize student enjoyment and achievement" (p. 3).

Sometimes referred to as kinaesthetic learning or somatic knowing, an embodied approach to learning connects body sensation to cognition, which is what happens when students become physically active in drama. Utilising the kinaesthetic languages of drama, including movement, sight, touch and sound, provides multiple access points, "so that children of different abilities might all find points of entry into the drama" (Nicholson, 2005, p. 57).

In drama, students learn through all their senses, through their body in movement, and in relation to others in the space. For example, learning science through drama can involve acting out scientific concepts such as lifecycles, magnetism, states of matter or atoms. These are sometimes referred to as physical simulations for learning science concepts (Abed, 2016). Students can work together to use imagination and create complex systems such as the human digestive system or a rainforest ecosystem, as described by Raphael and White (Chapter 'Transdisciplinarity: Science and Drama Education Developing Teachers for the Future'), using their bodies in space. Paige, Brown, O'Keefe, and Garret (Chapter 'Dramatising the S and the M in STEM') utilised this embodied approach to connect mathematical concepts in scientific learning such as enacting cogs, gears and simple machines. They refer to this as creative and body-based learning (CBL) and offer it as a pedagogical provocation in professional learning, to assist teachers to support students' conceptual learning and improve student engagement. In this kind of drama students are involved in cognition through physical activity, or 'thinking on their feet'. For some students this embodiment will be what is needed for lightbulb moments of understanding, for others it will help make the concepts more memorable, and for everyone it will be a relief from more sedentary modes of learning.

Drama as Social-Constructivist Approach to Science Learning

Drama is a social art form, sometimes called pro-social for the ways it brings participants together, often for creative group work and problem-solving (Neelands, 2010). Rather than being focused on individual success, group problem-solving in drama promotes the practice of teamwork, so necessary in science and many other fields of endeavour. As van Cuylenburg (Chapter 'The Science Drama Project: Meaning in the Middle') discovered in her science drama work with middle years students, the importance of interdisciplinary pedagogy in science and drama can sometimes become secondary to learning ways to learn with and about each other. James Davis (Chapter 'Dramatic and Undramatic Emotional Energy: Creating aesthetic and Emotive Learning Experiences in Science Classrooms') uses the term 'emotional energy', describing this as "the individual and collective experience of togetherness, social cohesion or solidarity arising through embodied performances of situated social practices". Siklander and Harmoinen (Chapter 'Ice Age is Approaching: Triggering University Students' Interest and Engagement in Gamified Outdoor Playful Learning Activities') found through located exploration within a prescribed scenario (a pending ice-age approaching), collaboration as members of clans was essential to generate learning and also resulted in engagement and enjoyment.

The pro-social nature of drama where diverse perspectives can contribute to the process of scientific inquiry or problem-solving is a student-centered and social-constructivist approach to learning in science (Çokadar & Yilmaz, 2010; Warner, 2013). Through the dialogic nature of drama, and the formation of a discourse community, groups of students are able to establish shared meanings (Braund et al., 2013).

Drama and Empathic Understanding for Socio-scientific Learning

The practice of drama involves exploring and celebrating the human experience. Not surprisingly, one of the most useful applications for drama in science is the potential for students to explore science in relation to wider human experience and concerns, including controversial issues in science and how scientific issues impact upon their lives and the lives of others (Dorion, 2009). Ever expanding scientific knowledge creates future opportunities but as the OECD remind us, "unprecedented innovation in science and technology, especially in bio-technology and artificial intelligence, is raising fundamental questions about what it is to be human" (2018, p. 3). Drama can put the human into scientific learning, and through taking on a variety of roles, students see the world through the eyes of another and come to understand and appreciate diverse perspectives. The aesthetics of the art of drama can allow for affective learning, a moment of profound realisation or a 'shock to thought' (Massumi, 2002; Thompson, 2009), which may lead to insight into the other, and empathy. This ability to develop empathy through the taking on of roles, including humans in diverse situations and non-human, is one of the unique affordances and the power of drama. It is of particular importance in understanding the complex ethical and social justice impacts of science on humans.

In some cases, the socio-constructivist and socio-scientific are intertwined as the process in the chapter by van Cuylenburg (Chapter 'The Science Drama Project: Meaning in the Middle') revealed. Fear and empathy were at the core of students' science-drama questions, and through exploring these cognitively they were able to enact them together in a contained environment. "The benefits of the science and drama *combination* were that it let students ask social-emotional questions under the security of scientific inquiry… and it enabled teachers and students to recognise and appreciate each other's capabilities in the broad skillset present across Science and drama" (van Cuylenburg, Chapter 'The Science Drama Project: Meaning in the Middle'). Similarly, Raphael and White in (Chapter "Transdisciplinarity: Science and Drama Education Developing Teachers for the Future') examine the ethics of stem-cell treatments through student research followed by taking on perspectives through roles in a whole class role-play, requiring debate and discussion in-role, followed by further discussion out of role. The combination of drama and science was symbiotic, through the different but intertwined ways each discipline applied pedagogical as well as contextual and conceptual framing.

Drama and Epistemic Knowledge-Thinking Like a Scientist

There are several examples in the chapters that make reference to the pioneering and widely influential drama work of Dorothy Heathcote (Bolton, 2002; Heathcote, 1991; Heathcote & Bolton, 1995; Wagner, 1979). Heathcote drew upon her experience of theatre to develop strategies for learning through drama such as Teacher in Role, Mantle of the Expert and Commission Drama, and for over fifty years these approaches have been developed continuously by educators for learning across the curriculum in a form often referred to as 'process drama' (O'Neill, 1995; Bowell & Heap, 2001). A Mantle of the Expert approach recasts students in the role of 'expert inquirers' committed to group construction of knowledge. This drama method provides a context that keenly engages students in the inquiry process (Warner, 2013, 2015, p. 261).

Encouraging students to think like and see themselves as scientists is a challenge that teachers face, and one they find can be met by dramatic pedagogies such as Mantle of the Expert. Swanson (Chapter 'Does Being Positioned in an Expert Scientist Role Enhance 11–13 Year-Old Students' Perceptions of Themselves as Scientists?') used Mantle of the Expert to position young students as scientists commissioned to investigate the sinking of the Wahine, encouraging students to think like scientists as they solve the mystery, and imagine themselves becoming scientists. Myers (Chapter 'New Eden: Mediating Pre-service Teachers' Conceptions of Education for Sustainable Development') uses multiple iterations of Mantle of the Expert to cast pre-service teachers in diverse roles so as to think through the economic, environmental and social aspects of sustainable development. Other dramatic pedagogies can also support students in thinking like a scientist, such as Baskerville and Anderson's (Chapter 'Responding to Climate Change: Developing Primary Children's Capability to Engage with Science Through Drama') use of drama prop and convention to engage students with the 'nature of science'. Evident in all of these approaches is opportunity for meta-cognition, thinking about thinking.

Drama for Communicating Science and Science Stories

Stories of scientists and their discoveries have long held a place in the nexus of science and drama—from Ibsen's (1882/1964) *Enemy of the People*, Dürrenmatt's (1964) *The Physicists*, or Brecht's (1938/1986) *Life of Galileo*, through to more modern explorations like Stoppard's (1993) *Arcadia*. These stories allow an audience a window into science and the lives of scientists, a place where empathy and insight can take hold. Drama can also put language to a seemingly inaccessible idea, as in Nicholas and Ng's (2008) development of *Hectic Electric* to explore electricity, its discovery, and its generation. Stories of science in education can peel back the inner layers of scientists' lives and practice (Ødegaard, 2003) and have been examined here through Sallis and Bird's (Chapter 'Australian Women in Science: A Model for a Research-Based Theatre Project in Secondary School Classrooms') development of research-based theatre for science stories, particularly focused on raising awareness of women in science. McGregor's (Chapter 'Stories from History: More Authentic Ways of Thinking Through Acting and Talking About Science') provides several examples of how the lives and practices of key scientists have been described via monologue to scaffold student learning towards understanding what motivates scientists to innovate and for designing their own investigations into how to make the 'best' plasticine. This collection was written during a time of the COVID-19 global pandemic, a time when science has loomed large in the news and science history is being made. Sue Davis (Chapter 'The Treatment of Dr. Lister: Investigating and Revisiting a Famous Primary Science-Based Drama') re-visits and analyses a well-known process drama, 'The Treatment of Dr. Lister', finding new relevance in a contemporary context. Students of today who experience this drama will be even more aware of the significance of Joseph Lister's 1865 discoveries in informing infection control, when we are aware there is still so much more we need to learn.

How Does Drama Benefit from Science?

The long list of benefits that drama can offer to learning in science might make it look as though science has more to gain in this symbiotic relationship. However, there are many important ways that drama also benefits from this relationship. Drama is not usually made about 'drama', rather, the art form is a way of exploring stories, themes and ideas that are of importance to humanity. The world of science has a rich history, and offers so many challenging, contemporary, topical, and ethical issues to explore as subject matter. The benefit here is that by engaging in science we can explore important content and issues, potentially creating very interesting drama, which if taken to the level of presentation or performance can be relevant, compelling and educative for wider audiences. Examples of this are found in Sallis and Bird's research-based performances, where participants learn about theatre-making processes while focusing on stories of successful women scientists (Chapter 'Australian Women in Science: A Model for a Research-Based Theatre Project in Secondary School Classrooms'), and Hannigan and Ferguson's endangered animal puppet 'theatre in a suitcase' created by secondary students for primary school audiences (Chapter 'Art-Science Education in the Anthropocene: Embodied Metaphor with Puppets and Performance').

Working with people practising science, drama educators and their students can benefit from invigorating collaborations, as seen in Raphael and White's chapter (Chapter 'Transdisciplinarity: Science and Drama Education Developing Teachers for the Future'). Drama is increasingly valued and drama educators respected for the particular knowledge and skills that they hold. This is important, as unfortunately in many educational institutions there is a disparity in the status of drama and science as subjects. STEM subjects frequently hold an importance in the eyes of students, parents, teachers, institutions and governments who mandate curriculum. This position of priority influences the amount of opportunity, time, money, resources and staffing given to the subject, as well as educational research opportunities and grants they are able to attract. When we demonstrate and model drama as the art form that liberates and deepens learning as well as one that communicates our understandings about the world for others, we promote drama as a discipline area, and the raising of its status. The drama science collaboration may also provide a gateway to studying drama as a discipline area for students and encourage new audiences who might otherwise not be likely to engage with drama or theatre. It may even prompt a greater appreciation of the role and impact of drama more broadly.

The value of drama in relation to science should not detract from the argument that drama, as an art form must maintain a place of its own in the curriculum. Drama has rich history, traditions, skills, techniques and processes of its own, and skills in drama feed into a valuable theatre and the broader arts industry. Although drama as an art form has been instrumental in developing human understanding since the dawn of time, and is extremely useful in this way, it must not merely be regarded as a pedagogical tool in service to other areas. To extend our metaphor of symbiosis, drama together with science is not *obligate mutualism*, where the survival of one or both is dependent upon the relationship. This is *facultative mutualism*, where both can benefit from but are not dependent upon the other (ThoughtCo., 2021). Just as the clownfish and sea anemone nourish and protect each other, drama and science both have something valuable to gain through their relationship with one another.

Intertwined Synergetic Model

In reading the science drama projects in the chapters in this collection, we come to see the pedagogy as a kind of intertwined synergy and we depict this in the model (Fig. 4). These examples of science drama involve separate disciplines that are entwined together like two bands of colour, sometimes they are overlapping, creating a new colour, and sometimes one or the other discipline colour comes to the fore in the rhythm of learning. We understand that at times an emphasis on one

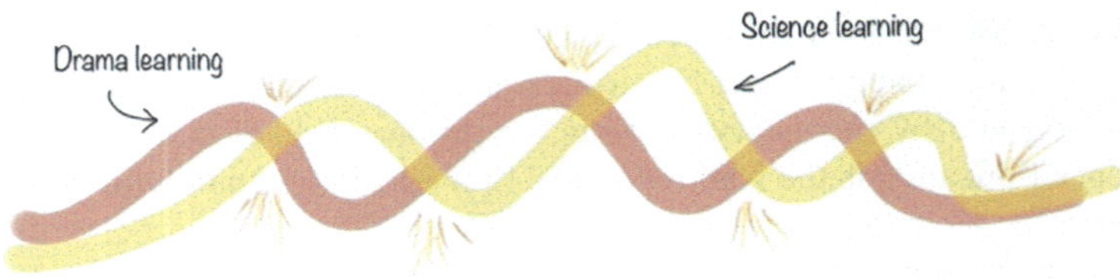

Fig. 4 The Intertwined synergetic model illustrating the movement of science and drama and the 'sparking' of new knowledge for students when the disciplines come together

discipline is what is needed. A synergy occurs when drama and science overlap and interact. Combined, they produce an effect that is often greater than the sum of their separate effects and we understand these moments of affect can be powerful and electric–sparking learning.

At the core of combining science and drama is the need for educators to spark learning in students. Every teacher knows the 'aha!' moment when they see it, and those sparks show up for each student in different ways. What we propose here is an intertwined synergetic model, where the process of combining drama and science opens up the potential for those sparks–either in science, drama or something that transcends both.

Recognising Tytler et al. (2019) work on temporal, intertwined interactions of the STEM disciplines, we also add a temporal layer to our model (Fig. 5), because we understand that this learning can be in the micro (minutes or hours), in the meso (a theme, topic or unit of work), or in the macro (a longer-term trajectory). These sparks may ignite in the micro-moments as Baskerville and Anderson (Chapter 'Responding to Climate Change: Developing Primary Children's Capability to Engage with Science Through Drama') noted when students made in-the-moment inferences, or in the meso where in Swanson's (Chapter 'Does Being Positioned in an Expert Scientist Role Enhance 11–13 Year-Old Students' Perceptions of Themselves as Scientists?') summative data highlights the learning over a term unit, or in the macro as van Cuylenburg (Chapter 'The Science Drama Project: Meaning in the Middle') found, sparks were generated a year after the

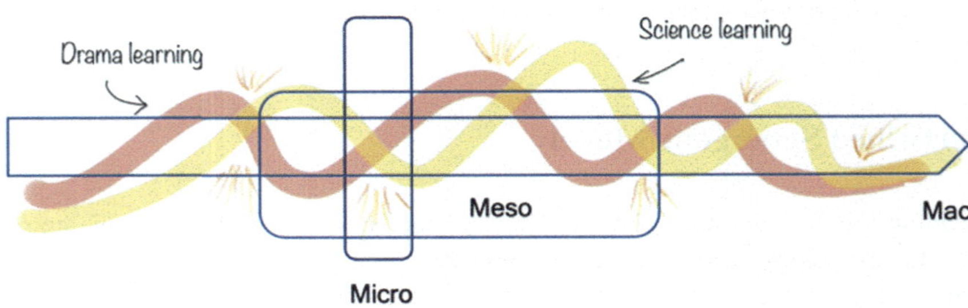

Fig. 5 The intertwined synergetic model illustrating temporality–learning in science/drama occurs over different periods of time

science drama project. The drama may facilitate science learning, the science may facilitate drama engagement, or the combination may yield something else entirely.

A temporal model, by focusing our attention on a specific time frame, helps us to appreciate what can be reasonably achieved within a given period of time, assisting with planning, teaching and evaluating science drama. A temporal view is also helpful when thinking about disciplinary integrity. It shows that a micro moment may simply focus on the learning in one discipline alone, and allows us to take a close-up view of what happens in those sparking moments of transdisciplinary learning not limited to one discipline. A meso view reveals the shifts between the disciplines over time, and the nuanced symbiotic relationships between them. A macro view can place the science drama experience within a broader context of learning and schools. We can view the pedagogical approach more generally and recognise other possibilities for learning when the approach is applied across and beyond the curriculum. Perhaps if drama was combined more often, one could develop knowledge and skills in students that they could apply in increasingly sophisticated ways.

Conclusion

In this chapter we have attempted to describe some of the reasons for the spark we see in bringing together drama and science. Alsop (2005) gathered contemporary theories about the role of feeling and emotion in science learning, explaining:

> Now, education is more than the memorisation of curriculum subject, the anaesthetised acquisition of a remote object. It is the beauty and delight of becoming absorbed, seeing the world in different ways with different possibilities. It is about challenge, surprise, desire, joy, expectation and mystery: the thrill of discovering oneself in relation to new ideas and contexts. (Alsop, 2005, p. 4)

None of the theories in Alsop's, 2005 volume specifically recognise drama as an approach, but since then, more has been learned about the potential for learning when drama and science work together. The examples in this collection show that drama provides opportunities for seeing the world in different ways and gives rise to challenge, surprise, desire, joy, expectation and mystery. Drama opens up and enhances opportunities for aesthetic learning, a welcome antidote for 'anaesthetising' teaching approaches that may dull students' senses. Many of these chapters reveal the rich potential for the aesthetic and the power of beauty in education (Winston, 2010), a welcome counter to utilitarian values, and a curriculum driven by narrowly conceived outcomes.

In this chapter, we focus on those sparks and shine the light of drama and science through a prism to reveal the many colours and qualities within a transdisciplinary approach. We note and sketch out models of science and drama. We expand the notion of a symbiotic relationship between drama and science, and suggest the metaphor of a mutual symbiotic relationship, as both science and drama have something to gain in relationship with the other. We also propose a model of intertwined synergy when drama and science are brought together. Overlaying a temporal consideration, we recognise that science and drama learning can be seen to occur in classes at the micro, meso and macro level. If any of the pedagogical models and metaphors presented here help others to think about their science drama practice, then they have served their purpose of moving the field towards a more theorised pedagogy of science and drama.

When we invited contributions of research chapters for this book, we were delighted with the number of educators and researchers who came forward enthusiastic about sharing their work. The majority of these chapters are the result of productive collaborations between colleagues who have made important discoveries through practice. They are all making valuable contributions to the formation of a theory of science drama pedagogy. Some chapters are specifically concerned with the important preparation of preservice teachers for creative and collaborative pedagogical approaches, and all chapters have something to offer to the practice of teachers in both science and drama. Most importantly, they are part of the ongoing conversation and collaboration that needs to continue across the discipline areas. We are living in times when creative interdisciplinary approaches are needed more than ever, and this volume provides examples of how this can be done. We hope that these chapters inspire, ignite and spark many more learning adventures in science and drama.

References

Abbs, P. (1989). *A is for aesthetic: Essays on creative and aesthetic education*. Psychology Press.

Abed, O. H. (2016). Drama-based science teaching and its effect on students' understanding of science concepts and their attitudes towards science learning. *International Education Studies, 9*(10), 163–173.

Abrahams, I., & Braund, M. (2012). *Performing science: Teaching physics chemistry and biology through drama*. Continuum.

ACARA. (2021). General capabilities in the Australian curriculum. Retrieved from https://www.australiancurriculum.edu.au/f-10-curriculum/general-capabilities/.

Akkerman, S. F., & Bakker, A. (2011). Boundary crossing and boundary objects. *Review of Educational Research, 81*(2), 132–169. https://doi.org/10.3102/0034654311404435.

Alsop, S. (2005). *Beyond Cartesian dualism: Encountering affect in the teaching and learning of science*. Springer.

Australian Research Council. (2008). *Living knowledge: Indigenous knowledge in science education*. Retrieved from https://livingknowledge.anu.edu.au/html/educators/index.htm

Bolton, G. M. (2002). *Dorothy Heathcote's story: Biography of a remarkable drama teacher*. Trentham Books.

Bowell, P., & Heap, B. S. (2001). *Planning process drama*. David Fulton Publishers.

Braund, M., Ekron, C., & Moodley, T. (2013). Critical episodes in student teachers' science lessons using drama in grades 6 and 7. *African Journal of Research in Mathematics, Science and Technology Education, 17*(1–2), 4–13.

Brecht, B. (1938/1986). Life of Galileo. In J. Willett (trans.), *Life of Galileo/Bertolt Brecht*. London, UK: Methuen.

Byrne, E. P., & Mullally G. (2015). Seeing beyond silos: Transdisciplinary approaches to education as a means of addressing sustainability issues. In W. Leal Filho & S. Nesbit (Eds.), *New developments in engineering education for sustainable development. World sustainability series*. Cham, Switzerland: Springer. https://doi.org/10.1007/978-3-319-32933-8_3.

Caglioti, G. (2017). Art according to Albert Einstein. *Lettera Matematica, 5*, 49–53. https://doi.org/10.1007/s40329-017-0155-7.

Capra, F. (2014). *Learning from Leonardo: Decoding the notebooks of a genius* (1st ed.). Berrett-Koehler Publishers.

Çokadar, H., & Yılmaz, G. (2010). Teaching ecosystems and matter cycles with creative drama activities. *Journal of Science Education & Technology, 19*(1), 80–89. https://doi.org/10.1007/s10956-009-9181-3.

Dewey, J. (1934). *Art as experience*. Capricorn Press.

Dinham, J. (2019). *Delivering authentic arts education* (4th ed.). Cengage.

Dorion, K. (2009). Science through drama: A multicase exploration of the characteristics of drama activities used in secondary science classrooms. *International Journal of Science Education, 31*(16), 2247–2270.

Dunn, J., & Stinson, M. (2011). Not without the art!! The importance of teacher artistry when applying drama as pedagogy for additional language learning. *Research in Drama Education: THe Journal of Applied Theatre and Performance, 16*(4), 617–633. https://doi.org/10.1080/13569783.2011.617110.

Dürrenmatt, F. (1964). *The Physicists* (J. Kirkup, Trans.). New York: Grove Press.

Hannigan, S., Raphael, J., White, P., Bragg, L., & Cripps Clark, J. (2016). Collaborative reflective experiences and practice in education explored through self-study and arts-based research. *Creative Approaches to Research, 9*(1), 84–110.

Heathcote, D. (1991). *Collected writing on education and drama*. Northwestern University Press.

Heathcote, D., & Bolton, G. (1995). *Drama for learning: Dorothy Heathcote's mantle of the expert approach to education*. Heinemann.

Hornbrook, D. (1989). *Education and dramatic art*. Basil Blackwell.

Ibsen, H. (1882/1964). An enemy of the people. In P. Watts (Trans.), *Ghosts and other plays* (pp. 103–220). Penguin.

Kang, N.-H. (2019). A review of the effect of integrated STEM or STEAM (science, technology, engineering, arts, and mathematics) education in South Korea. *Asia-Pacific Science Education, 1*. https://doi.org/10.1186/s41029-019-0034-y.

Lerman, L. (2014). *Hiking the horizontal field notes from a choreographer*. Wesleyan University Press.

Mark, J. (2019). Hildegard of Bingen, ancient history Encyclopedia. Retrieved from https://www.ancient.eu/Hildegard_of_Bingen/.

Massumi, B. (Ed.). (2002). *A shock to thought: Expressions after Deleuze and Guattari*. Routledge.

Menken, M., & Keestra, S. (2016). *An introduction to interdisciplinary research: Theory and practice*. Amsterdam, The Netherlands: Amsterdam University Press. Retrieved from https://philarchive.org/archive/MENAIT-4.

Merriam-Webster. (2021). Symbiosis. Retrieved from https://www.merriam-webster.com/dictionary/symbiosis.

Neelands, J. (2010). The art of togetherness. In P. O'Connor (Ed.), *Creating democratic citizenship through drama education: The writings of Jonothan Neelands* (pp. 131–142). Trentham Books.

Nicholas, H., & Ng, W. (2008). Blending creativity, science and drama. *Gifted & Talented International, 23*(1), 51–60.

Nicholson, H. (2005). *Applied drama: The gift of theatre*. Basingstoke: Palgave Macmillan.

Nicholson, H. (2011). *Theatre, education and performance: The map and the story*. Palgrave Macmillan.

Ødegaard, M. (2003). Dramatic science. A critical review of drama in science education. *Studies in Science Education, 39*(1), 75–101.

OECD. (2018). The future we want. *The future of education and skills 2030*. Retrieved from https://www.oecd.org/education/2030/E2030%20Position%20Paper%20(05.04.2018).pdf.
OECD. (2019). Knowledge for 2030. *The future of education and skills 2030*. Retrieved from https://www.oecd.org/education/2030-project/teaching-and-learning/learning/knowledge/Knowledge_for_2030_concept_note.pdf.
O'Neill, C. (1995). *Drama worlds: A framework for process drama*. Heinemann.
O'Toole, J. (2017). The arts, creativity & play. In C. Sinclair, N. Jeanneret, J. O'Toole, & N. A. Hunter (Eds.), *Education in the arts* (3rd ed., pp. 17–30). Oxford University Press.
Raphael, J., Hannigan, S., & White, P. (2016). Drawing out understandings of collaborative self-study in teacher education. In D. Garbett & A. Ovens. (Eds.), *Enacting self-study as methodology for professional inquiry* (pp. 109–118). Retrieved from https://www.castleconference.com/conference-history.html.
Slade, P. (1954). *Child drama*. London, UK: Hodder & Stoughton.
Snow, C. P. (1959). *The two cultures*. Cambridge University Press.
Stoppard, T. (1993). *Arcadia: A play in two acts*. Concord Theatricals.
Thompson, J. (2009). Performance affects: Applied theatre and the end of effect. Basingstoke, England & New York: Palgrave Macmillan.
ThoughtCo. (2021). Mutualism: Symbiotic relationships. Retrieved from: https://www.thoughtco.com/mutualism-symbiotic-relationships-4109634.
Turkka, J., Haatainen, O., & Aksela, M. (2017). Integrating art into science education: A survey of science teachers' practices. *International Journal of Science Education, 39*(10), 1403.
Tytler, R., Prain, V., & Hobbs, L. (2019). Re-conceptualising interdisciplinarity in STEM through a temporal model. *Research in Science Education*. https://doi.org/10.1007/s11165-019-09872-2.
UNESCO. (n.d). Glossary of curriculum terminology, interdisciplinary approach. Retrieved from http://www.ibe.unesco.org/en/glossary-curriculum-terminology/t/transdisciplinary-approach.
Vygotsky, L. S. (1976). Play and its role in the mental development of the child. In J. S. Bruner, A. Jolly, & K. Sylva (Eds.), *Play: Its role in development and evolution* (pp. 536–554). Penguin.
Wagner, B. J. (1979). *Dorothy Heathcote: Drama as a learning medium*. Cheltenham, UK: Stanley Thornes.
Wapnick, E. (2015). *Why some of us don't have one true calling* [Streaming video]. TEDxBend. Retrieved from https://www.ted.com/talks/emilie_wapnick_why_some_of_us_don_t_have_one_true_calling#t-34903.
Warner, C. D. (2013/2015). Drama and science: An unlikely partnership for inquiry. In M. Anderson & J. Dunn (Eds.), *How drama activates learning: Contemporary research and practice* (pp. 260–276). Bloomsbury Academic.
Westley, F. R., & Folke, C. (2018). Iconic images, symbols, and archetypes: Their function in art and science. *Ecology and Society, 23*(4), 31. https://doi.org/10.5751/ES-10495-230431.
Winston, J. (2010). *Beauty in education*. Routledge.
White, P. J., Raphael, J., Hannigan, S., & Cripps Clark, J. (2020). Collaborative reflective experience and practice in education: A model for a self-study community of practice. *Australian Journal of Teacher Education, 45*(8). Retrieved from https://ro.ecu.edu.au/ajte/vol45/iss8/6/.

Jo Raphael (B.Ed., M.Ed., Ph.D., SFHEA) is senior lecturer in drama education at Deakin University. She teaches in postgraduate and undergraduate pre-service teacher education programs and remains active in school and community settings. She has applied drama for learning within diverse contexts including museums and galleries and in areas of the humanities, languages, science and environmental sustainability. Jo is Artistic Director of Fusion Theatre, an inclusive community-based theatre company. She has won multiple awards for her teaching and has been awarded for her extensive contributions to her professional community. Her research and publications span the areas of arts curriculum and pedagogy, teacher education, inclusive education and teaching for diversity.

Peta J White (B.Sc. (Hons), Dip.Ed, M Rur Sys Man, Ph.D., SFHEA) is a science and environmental education senior lecturer at Deakin University. Peta has worked in classrooms, as a curriculum consultant and manager, and as a teacher educator in several jurisdictions across Canada and Australia. Peta gained her Ph.D. in Saskatchewan, Canada where she focussed on learning to live sustainably which became a platform from which to educate future teachers. Her passion for initial teacher educator, environmental education/academic activist work, and action-orientated methodologies drives her current teaching/research scholarship. Peta's current research interests follows three directions including science and biology education; sustainability, climate change, and environmental education; and collaborative/activist research.

Kitty van Cuylenburg (B.A./B.Sc. (Hons), M.Teach.) is a VIT registered drama and science teacher, a Teaching and Leadership advisor for Teach for Australia, and theatre maker and board member of In The Park Productions. She has a background in mine-site hydrogeology, has worked for an education startup in NYC, performed in and produced theatre in Melbourne (MICF, MIFF), and has a passion for interdisciplinary practice of science and drama. She is an erstwhile research student at Deakin University, and undertakes action research in this space through her teaching practice.

Mutuality and Inter-relativity of Drama and Science

Tricia Clark-Fookes and Senka Henderson

Abstract Drama and science are often considered as pedagogical binaries; one concerned with rationality and the other affect. This chapter will explore the notion of mutuality and the inter-relationships between the two disciplines. Neither discipline in service of the other, but one enriching the other. This positions the two disciplines as mutual and symbiotic. Demonstrated in this chapter is a theoretical approach to interdisciplinary unit design that draws on an understanding of the mutuality of the two disciplines, using the Mantle of the expert and 5Es inquiry model. The unit titled 'Inter-relationships in Our World' brings together the content from Australian Curriculum Assessment and Reporting Authority (ACARA) at Year 9 in both drama and science.

Keywords Drama · Science · Interdisciplinary · Mantle of the expert · 5Es inquiry model

Introduction

> (The) polarization [of art and science] is sheer loss to us all. To us as people, and to our society. It is at the same time practical and intellectual and creative loss, and I repeat that it is false to imagine that those three considerations are clearly separable. (Snow, 1959, p. 6)

Snow in his now famous 1959 Rede lecture asserted the existence of "two cultures": science and the arts. He perceived a divide between the two fields as a loss to society, arguing that both fields should build bridges and engage further to progress of human knowledge and benefit society. In Snow's view, the meeting of knowledge fields reaps "creative chances" that result in advances.

> The clashing point of two subjects, two disciplines, two cultures—of two galaxies, so far as that goes—ought to produce creative chances. In the history of mental activity that has been where some of the break-throughs came. (Snow, 1959, p. 8)

T. Clark-Fookes (✉) · S. Henderson
Creative Industries, Education and Social Justice Faculty, Queensland University of Technology, Brisbane, Australia
e-mail: tricia.clarkfookes@qut.edu.au

P. J White et al. (eds.), *Science and Drama: Contemporary and Creative Approaches to Teaching and Learning*, https://doi.org/10.1007/978-3-030-84401-1_2

Highlighted here is a view of science and art as mutually beneficial, generating new knowledge through the meeting of epistemic fields. In Snow's view art and science exist in polarities that can be bridged, a view shared by contemporary academics (Braund, 2015; Braund & Reiss, 2019; Najami et al., 2019). Similarly, academic and artist Liz Lerman disputes the conception of knowledge in polarities offering an alternative perspective termed "hiking the horizontal" (Lerman, 2014). In this view, knowledge exists along a spectrum of multiple perspectives seeking to "embrace multiple forms of knowledge and discovery" which may be united by a common purpose (Lerman, 2014, p. xvi). Integral to this conception of multiple perspectives is the rejection of hierarchies. "In a hierarchical order, the only way to make a distinction is to literally put the other idea "down". There is only room for one idea. But in hiking the horizontal, many ideas can coexist" (Lerman, 2014, p. xvi).

Lerman's suggestion conceptualises epistemic fields as co-existent and of equal value with practitioners shifting along the spectrum to take advantage of the various epistemic frames offering a richer perception of the concept at hand. 'Hiking the horizontal' offers a way of conceptualising the relationship between arts and science (ipso facto drama and science) that rejects polarities and hierarchies. Aesthetic theorist Willis (1998) holds a similar understanding, suggesting that engaging in "symbolic" and "creative" conceptions of theory can, "produce symbolic extension, a meaning-making which is more horizontal, more detached from the conditions of production" (Willis, 1998, p. 164). That is, by engaging with knowledge through different epistemological lenses and symbolic representations, learners can gain a more complete understanding of a concept. An integration of cognitive, affective and procedural knowledge takes place (Dorion, 2009; Ødegaard, 2003; Verhoeff, 2017).

In the neo-liberal agenda of the current educational climate here in Australia and abroad, science and STEM knowledge domains are privileged over Arts. Ewing's (2011) report into The Arts and Australian Education found that, "Mathematics and science are prioritised over humanities and the Arts" (p. 9). Similar claims are made in other regions of the world (Robinson & Aronica, 2015). For evidence one only need look at the time allocation provided for both learning areas across educational jurisdictions. If we subscribe to Snow's view of art and science as mutually beneficial then traditional dichotomies and boundaries of the academy need rethinking. Lerman's conception of 'hiking the horizontal' offers such a model for how this may be achieved (Lerman, 2014).

Traversing the Horizon of Drama and Science

Drama and science are distinct fields of knowledge, and in the educational context, each possesses its own unique pedagogical approaches. So how is it possible for the two fields to occupy space on the same horizontal plane of knowledge? Perhaps a helpful starting point for unifying drama and science is provided by Sousa and Pilecki (2018) who highlight that, "the main objective of both art and science is discovery" (p. 17). The mutual commitment to the principle of discovery drives both learning areas and defines the pedagogical approaches of both fields.

Discovery is a way for learners to move from the unknown to the known (Bruner, 1966), placing learners in an active position inquiring into concepts (Coffman, 2017; Hammer, 1997; Kidman & Casinader, 2017). Discovery as used here is synonymous with inquiry, a pedagogical approach employed in drama and science. Although each learning area employs different inquiry processes, the notion of discovery is the same. The driver for the discovery/inquiry learning in both disciplines is the unfinished nature of the inquiry with incomplete or missing information. Each discipline uses different process to inquire, but through the experience of seeking missing information, students are discovering for themselves the intended content. According to Montuori (2012) "the inquirer [student] is not a spectator to the world, but embodied and embedded, an active participant in knowledge-creation and praxis" (p. 342).

Snow (1959), as previously discussed, highlighted a conception of art and science as binaries suggesting that that two fields view each other's language as different and foreign. Science views its language use as rational and objective dealing with observable facts, whilst art, particularly drama provides insights into the world as described through aesthetic language expressing both physical and non-physical impressions of the world and experience. In our view, what is often missed is that both fields function to make sense of the world and its experiences through observation and curiosity; that is both prioritise meaning and sense making. We assert that observation is key to the knowledge generated by both fields. The divergence lies in how those observations are captured and communicated.

Drama and science function through symbol systems to present and communicate knowledge. Langer (1957) tells us that, "symbol-making function is one of man's primary activities, like eating, looking, or moving about" (p. 41). It is the fundamental process of the mind and goes on all the time. A symbol in this respect articulates and presents concepts (Langer, 1953). The symbol systems employed by drama and science differ in their form and function. Science draws primarily on discursive symbols systems, whilst drama primarily employs non-discursive symbol systems. Discursive symbol systems rely on words ("verbal symbolism") (Langer, 1957, p. 81), both written and spoken. In contrast, non-discursive symbol systems attempt to articulate "things which do not fit the grammatical scheme of expression" (Langer, 1957, p. 88). "Non-discursive symbolization is simply a term that accounts for the many other ways humans use symbols to create meaning—methods wholly outside the realm of traditional, word-based, discursive text" (Murray, 2009, p. 12).

Murray (2009) highlights the accepted privileging of discursive symbol systems, suggesting "[t]hough each type of symbolization is needed, useful, and important, [non-discursive] is the most neglected in many discussions about symbolization and language" (p. 1). Yet the two symbol systems communicate different understandings and knowledge through their distinct forms. Discursive knowledge adheres to the law of language where ideas are "strung out" (Langer, 1957, p. 81) side by side in order to create meaning. The temporal rules governing language necessitate the expression of ideas sequentially and inadequately captures the simultaneous and complex aspects of concepts. "Only thoughts which can be arranged in this peculiar order can be spoken at all; any idea which does not lend itself to this "projection" is ineffable, incommunicable by means of words" (Langer, 1957, p. 81).

In contrast, non-discursive symbols articulate knowledge that cannot be communicated via discursive means (Langer, 1953). Non-discursive symbol systems have the capability of articulating complex ideas that cannot be adequately expressed through discursive means. "They are both syntactically and semantically "dense" [original emphasis]" (Salomon, 1979, p. 133). Murray (2009) asserts it is a "symbolized language" with the capacity to accommodate meaning to the "unutterable, affective, ephemeral" (p. 5). Discursive symbol systems as previously highlighted operate through a sequential arrangement of concepts, ruled by the laws of language. In contrast, non-discursive systems are not limited to the chain-of-reasoning we require in discursive text" (Murray, 2009, p. 5).

> They do not present their constituents successively, but simultaneously [...] Their complexity, consequently, is not limited, as the complexity of discourse is limited, by what the mind can retain from the beginning of an apperceptive act to the end of it. [...] An idea that contains too many closely related parts, too many relations within relations, cannot be "projected" into discursive form; it is too subtle for speech. (Langer, 1957, p. 93)

Discursive and Non-discursive text thus offer different understandings and can illuminate different aspects of a phenomena using their unique means. In academic settings, great value has been ascribed to discursive symbol systems, which suggests a hierarchy. This privileging of the discursive has limited our capacity to engage with complex concepts that need to be captured in simultaneous articulations rather than sequential articulations as discursive forms take. In addition, the discursive limits our capacity to express the "unutterable, affective, ephemeral" (Murray, 2009, p. 5), thus limiting our understanding of concepts. The non-discursive is not an alternative means of expression, it is an additional means that enriches our understanding of phenomena. In current conceptions of the curricular the non-discursive is viewed as supplementary and an unnecessary distraction from core business. This hierarchical view fails to acknowledge the depth of insight gained when the best of both forms of symbol systems are employed. What is required is a model of learning that embraces the value of both discursive and non-discursive forms, placing them side by side, hiking the horizontal to achieve quality outcomes for learners.

This horizontal conception of curriculum using discursive and non-discursive means is exemplified by the interdisciplinary pairing of drama and science. Science traditionally employs discursive symbol systems to communicate its understandings, yet it recognises and places value on experience as evidenced in the experimental inquiries performed by its students.

Science education values and embraces the experience of scientific inquiry. This centrality of experience aligns to Dewey's (1986) assertions of the "intimate and necessary relation between the processes of actual experience and education" (p. 242). The inclusion and thus valuing of experience in Science education points to an acknowledgement of knowledge generated through haptic and tacit means, yet the academy demands these understandings be communicated via discursive means which may not best capture the full experience.

Similarly, drama education prioritises experiential learning whilst allowing students to articulate the breadth of their knowledge using both discursive and non-discursive symbol systems. Dramatic form allows students to navigate affective, objective, sensorial, haptic and linguistic understandings of concepts by fusing language and embodied means of communication. O'Toole and O'Mara highlight this protean quality of drama education.

> Drama is chameleon like, drawing on multiple artforms in its practice whilst serving an array of purposes, it has "the ability to change shape at will and accomplish a wide diversity of quite different ends…. a wonderful and almost magical capacity. (O'Toole & O'Mara, 2007, p. 127)

The Symbiosis of Drama and Science in the Development of a Unit of Work

As established, we consider drama and science to not only share some common values and approaches to learning, but we believe that when brought together they create a symbiotic relationship, mutually benefiting each other. Science and its emphasis on rationality, empirical inquiry and discursive expression is aided by the affective understandings and non-discursive representation of ideas in drama, and vice versa. Together, the two disciplines offer a richer and more complete understanding of concepts. In developing an interdisciplinary unit of work, we exemplify the mutuality and inter-relationship between the two disciplines. Neither discipline functions in service of the other, instead our approach to unit design frames each discipline as enriching the other. We consider each discipline as more than a distinct field of knowledge, but as separate symbol systems through which we can communicate knowledge. Each symbol system can be used to explore the same content but offer different viewpoints and insights. We build on Langer's (1957) views that increased understanding and knowledge can be experienced when we engage both discursive and non-discursive symbol systems to transform abstract knowledge and connect it to affective and sensorial or embodied understandings.

By framing the two disciplines as symbols systems, subject matter can be explored through a common context. This context unifies students' experience and understanding whilst providing an authentic encounter with the knowledge and processes of both drama and science. This belief underpins the design of an interdisciplinary unit of work serving both Year 9 science and Year 9 drama content in the Australian Curriculum. Specifically, this unit of work examines the "Inter-relationships in our World". The unit is designed as a case study examining bees and their relationship to the ecosystem and human activity. Through the context of bees, students will explore content descriptors from ACARA's Year 9 science and Year 9 drama curriculum (see Table 1).

Table 1 The year 9 ACARA content descriptors

Discipline	Content Descriptors
Science	• Ecosystems consist of communities of interdependent organisms and abiotic components of the environment; matter and energy flow through these systems (ACSSU176) • People use scientific knowledge to evaluate whether they accept claims, explanations or predictions, and advances in science can affect people's lives, including generating new career opportunities (ACSHE160) • Critically analyse the validity of information in primary and secondary sources and evaluate the approaches used to solve problems (ACSIS172) • Communicate scientific ideas and information for a particular purpose, including constructing evidence-based arguments and using appropriate scientific language, conventions, and representations (ACSIS174)
Drama	• Improvise with the elements of drama and narrative structure to develop ideas, and explore subtext to shape devised and scripted drama (ACADRM047) • Perform devised and scripted drama making deliberate artistic choices and shaping design elements to unify dramatic meaning for an audience (ACADRM051)

The selected science content challenges students to examine ecosystems and the inter-relationships among species, and environmental factors in these complex systems. The drama content of the unit makes metaphoric and literal connections between environmental inter-relationships and the dramatic elements of relationship and tension. More specifically, tension of relationship is used to frame students' understanding of the tensions and mutualities in ecosystems. Here we see the elements of drama and the scientific concept of inter-relationship operating mutually to assist students' development of knowledge. The subject matter of each discipline area is unified by the context and common exploration of bees and their relationship to the ecosystem. The figure below outlines the relationship between the subject matter of the two disciplines and their shared context (see Fig. 1).

The unit challenges students to synthesise the knowledge and understanding of the two disciplines into a single assessment outcome; a multimodal performance. This single assessment item asks students to convey their understanding via discursive and non-discursive forms. That is, they communicate scientific evidence and understandings with the affective and sensory methods of dramatic form. Here in this single assessment item, we gather data regarding students' achievement in both science and drama. We have selected this unified approach to assessment for two purposes:

1. to offer students different ways of communicating their knowledge and understanding, and
2. reducing the assessment load by combining assessment for two disciplines into a single task.

The assessment task culminates in a multimodal performance. This performance outcome is the assessable product generated through a project-based approach to learning. Project-based learning involves the learner in active construction of concepts, facts and principles through carrying out investigation in collaboration with others, while creating an authentic artefact that represents student understanding

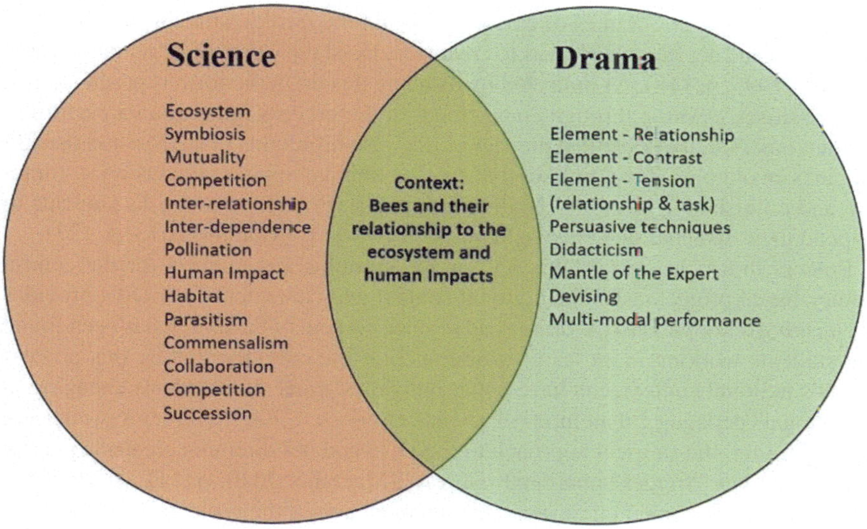

Fig. 1 The relationship between the subject matter for Science and Drama

(Crawford, 2014; Marx et al., 1997). This pedagogical approach exploits the experiential and inquiry-based approaches valued in both science and drama. The project-based approach offers students the opportunity to engage in authentic learning regarding the importance of bees in ecosystems and how humans have impacted bees and as result the balance of the ecosystem. This learning context challenges students to engage with a present real-world concern that has the potential to impact their life and the world they inhabit. Connecting learning to lived contexts and common culture allows students to ground theoretical knowledge and connect knowledge to everyday life. It has been shown that learners have a tendency to separate "school knowledge from everyday application" (Andersen, 2004, p. 283). Paul Willis terms this locating of knowledge in "common culture" or "lived culture" as grounded aesthetics. That is, he asserts the value of emotional "appropriation" of content "to produce *new* connections and meanings" for learners (Willis, 1998, p. 164).

In keeping with Willis' concept of grounded aesthetics drawing on common culture as a point of access and engagement for students, the central project-based inquiry of the unit (bees and their importance in the ecosystem) engages students in a familiar and somewhat satirical fictional context. The context of the inquiry project is driven by a celebrity influencer character, The Diva. She is framed as a parody of the celebrity influencers students are familiar with from social media including their ill-informed and tempestuous rants on social media. A caricature of a social media 'bridezilla' provides the central inquiry of the unit, as she poses a challenge to students to "save her wedding day".

Students are introduced to the Diva and her challenge through a process drama. Process drama is a drama-based form and pedagogical approach that engages students in a series of dramatic frames or episodes. Within the frames, students and teachers "improvise and act [together] so as to create and build the world of a dramatic event" (O'Neill, 1995, p. 118). As indicated by its name this dramatic form is predicated on and prioritises learning through engagement in the process rather than a performative outcome. Cecily O'Neill, a pioneer of the form highlights that, "*Process* usually indicates an ongoing event, unlike a *product*, a term that implies conclusion, completion and a finished "object" (O'Neill, 1995, p. xv). Process drama asks students to suspend their disbelief and engage in an 'as if' world (O'Connor, 2013, p. 127).

Process drama is used in this unit as the stimulus and context for the central inquiry-based project task. The fictional context and character of the Diva provides an engaging entry point to the drama and science content of the unit. It has been found that students working in an 'as if' world such as the one provided by this process drama's fictional context, can have better outcomes/gains than students engaging in "traditional" pedagogical methods of science (Andersen, 2004; Warner & Andersen, 2004). Further, the satirical approach to the Diva and her dilemma creates a playful context that can "trigger situational interest" (Marzano, 2010, p. 11), which is the foundation for sustained interest and engagement across the project.

The 'as if' we are asking students to engage with involves a stereotypical Diva and social media 'influencer'. The Diva shares her life (and tantrums) through her social media posts which are followed by millions of adoring fans. The premise of the drama involves the Diva, planning her wedding when she realises that her dream wedding is not achievable because there are no flowers and no fruit available for her grand wedding vision. Being a Diva, she offers huge sums of money for anyone who can solve the issue and make sure her wedding day has bountiful flowers and fruit.

The process drama serves as a unifying device for the two curriculum areas. The scientific inquiry is established whilst students engage in exploration of dramatic form and the identified content descriptors. Students are provided with a fictional context and stimulus to drive their imaginative and scientific inquiry using process drama form. Process drama and inquiry share a defining quality of being unfinished. The unfinished nature of the process drama, requiring students to complete the narrative aligns with the incomplete or missing information that drives inquiry processes. The missing information in both the process drama and the project-based inquiry of this unit appeal to the human tendency to attend to situations that have missing details and attempt to fill in the blanks (Ebbinghaus, 1913) and "make sense of the world" (Loewenstein, 1994, p. 82). The process drama acts as a launch pad for the scientific inquiry by presenting missing information whilst offering an aesthetic approach to the scientific content. This mode of inquiry seeks to motivate the learner and values the integrated knowledge generated through "experience, affect, and subjectivity in the learning process" (Montouri, 2012, p. 834). That is, it engages students cognitively, affectively and perceptively which reflects an aesthetic pedagogical approach.

Aesthetic education theorist Abbs (1987) suggests aesthetic experiences are "simultaneously perceptive, affective and cognitive [and] offer an education, therefore, of the highest order not through the analytical intellect but through engaged sensibility" (p. 55). That is, rather than creating learning experiences that cater to the cognitive only, as frequently occurs in science education, students are immersed and engaged in affective understandings and perceive knowledge through a range of senses. Using learning experiences that engage students cognitively and through the senses is common in science, but the inclusion of affective understandings expands the pedagogical approach to engage students' emotions and personal interests in the learning. Carnegie Hall's education director Sarah Johnson (cited in Booth, 2012, para. 14) asserts that the ultimate outcome of aesthetic engagement is "intrinsically-motivated curiosity". In this instance, students' emotions and interests are engaged through the fictional context with the aim of fostering a sense of, "animation, connection and heightened awareness" (Bundy, 2005, p. 2). Connections are continuously made between the learners' world and content across the unit of work.

The central inquiry is directly connected to the function and importance of bees in our ecosystem. Throughout the unit, students are asked to imagine what the world would be like without bees and consider the impacts of this on their own lives. Students are provided with articles and other resources that provide both empirical and subjective accounts of the endangered nature of bees in current times. Jointly, these artefacts allow students to develop a fulsome understanding of not just facts, but the affective impacts which serve to provide intrinsic motivation for the students' inquiry process (Andersen, 2004).

The unit task is designed as a collaborative, guided inquiry, where students are provided with the problem, however they must devise their own methods to reach a conclusion. To engage in the inquiry process, students are enrolled in groups as scientists, government officials and farmers to address the Diva's challenge using the 5Es inquiry instructional model (Bybee, 1997, 2015). Consisting of 5 distinct, but related phases; Engage, Explore, Explain, Elaborate and Evaluate, the teaching strategies used throughout the unit are designed to motivate active learning (Bunterm et al., 2014; Bybee, 2015). By planning for students to complete the task in small groups, a learning environment is created which fosters the social construction of knowledge as students collaborate with their peers to solve the inquiry-based problem of the task (Oliver & Venville, 2012). Students move through the 5 phases of inquiry tackling the problem at hand from the viewpoint of scientists, government officials or farmers. This mode of the inquiry is a drama-based pedagogy called Mantle of the expert (Heathcote, 1995; Heathcote & Herbert, 1985). Mantle of the expert enrols students in a fictional context within which they are endowed with expertise (Heathcote & Herbert, 1985). Using this expertise, they respond 'as if' they are 'experts' providing expert advice to complete a project. In the case of this unit, they are endowed as experts who engage in research to solve the Diva's wedding dilemma. In small groups students work as colleagues and collaborate to solve this fictional dilemma using their expert knowledge of bees and the ecosystem.

As students deal with the Diva's dilemma, they engage with scientific content and work in role. Drama skills are used across the unit as both a context for science-based learning content, and dramatic elements and form will serve as the content to meet the drama curriculum outcomes. The sign systems and ways of working of the two discipline areas blend to culminate in the final project task where students present their solution to the Diva. In a performance, students fuse their understanding of ecosystems, the significance of bees in the ecosystem and human impacts on bees in ecosystems, with dramatic form using persuasive dramatic techniques and communicating a didactic viewpoint. That is, they will create a final performance persuading the Diva that she should support their group's plan to re-balance the ecosystem and ensure bees' role as pollinators is understood and protected.

"Inter-relationships in Our World": Locating the Mutuality Across an Interdisciplinary Unit of Work

"Inter-relationships in our World" is an interdisciplinary unit of work. Here the term interdisciplinary refers to "any form of dialogue or interaction between two or more disciplines" (Lindvig & Ulriksen, 2020, p. x). We have made connections beyond a theme or issue (Venville et al., 2000) to examine the commonalities embedded in the two disciplines. Both disciplines work collaboratively to "advance fundamental understanding beyond the scope of a single discipline or area" (Sciences, Engineering, & Medicine, 2005, p. 2).

In designing the interdisciplinary unit, the notion of symbiosis provides both a metaphoric structure (the mutuality of science and drama) and models the subject matter of symbiosis explored in the unit through its interdisciplinary practices. Across the unit structure whilst addressing the 5Es inquiry phases, the two disciplines organically come together and separate at points where the subject matter in focus is best served. That is, at points where knowledge is best understood through the fusion of the discipline areas, we see the subject matter explored through both the science and drama symbol systems and ways of working simultaneously. At some points in the unit, subject matter is best served through the lens of a single discipline to deepen aspects of students' understanding. This is will be explained in relation to the unit plan below (see Table 2).

Drawing on the 5Es, Engage, Elaborate and Evaluate are experienced with both disciplines combined, whilst Explore and Explain are divided between drama and science. The Explore phase has been experienced through the drama so as to capitalise on students' affective engagement with the issue of bees' and their role in the ecosystem and relationship to human activity. The Explain inquiry phase is delivered through the discipline and subject matter of science only.

Table 2 Unit outline—'inter-relationships in our World'

Week No.	Underpinning pedagogies	Learning content and sample experiences	
		Science	Drama
1	5Es—engage Aesthetic—simultaneously cognitive, affective and sensory	Exploring ecosystem concepts of inter-relationship, symbiosis, mutualism commensalism, parasitism, collaboration, competition and succession through drama games • e.g. Freeze tag game that reveals ecosystem with flowers, pollinator and human. Then variations on game to metaphorically explore symbiosis and human impact on ecosystem • e.g. group sit and other balance challenges as metaphors for inter-relationship • e.g. Re-work 'Bomb and shield' game as metaphor for ecosystem inter-relationships (including predator, symbiont and competition) • e.g. Balloon Passing challenge as metaphor for Succession. Students are challenged to pass a balloon across group. As the challenge progresses the teacher adds further accumulating challenges by constraining the use of parts of the body. The group must complete the balloon passing task by adapting to the changes • e.g. Drama roleplays exploring the element of relationship and the tension of relationships as metaphors for inter-relationships in ecosystems and the impacts of change	
2	5Es—engage Aesthetic—simultaneously cognitive, affective and sensory	Process drama: Diva wedding *Frame 1*: Diva announces her engagement on Instagram *Frame 2* (belief building): rumour mill: In small groups students invent a story about an incident involving the celeb acting as a total diva *Frame 2* (injection of tension): the Diva releases a press release outraged that her wedding is going to be ruined because of some environmental issue. Photo of her wedding as envisioned (no flowers—balloon bouquet, bread, no cake, no fruit or vegetables, she will have to eat carbs!! shock horror). She offers a reward for anyone who can ensure her wedding has flowers, fruit for the cake and a vegan feast of fresh fruit and vegetables *Frame 3*: town hall meeting. Diva's wedding planner brings the leading officials together for an initial think tank. Wedding planner welcomes the world's leading scientists, government officials and farmers and asking the questions: "What could have caused this issue?" "How can we solve the problem?" *Frame 4*: enrolment and spilt into groups of 3–4 (world's leading scientists, government officials and farmers). Students create scenes answering the questions: "How did we get dragged into this? What has this got do with us?"	

(continued)

Table 2 (continued)

Week No.	Underpinning pedagogies	Learning content and sample experiences	
		Science	Drama
		Frame 5: setup project-based task. Create a pitch for the Diva telling her how you are going to solve the problem. You will need to ensure that you have a sound scientific base, but also ensure that the presentation is engaging as the Diva has a short attention span. You want to get the money!!	
3	5Es—explore Aesthetic—simultaneously cognitive, affective and sensory	Ecosystem concepts View Bee movie scenes and discuss • Explanation of pollination (approx. 16 min) • Effects of stopping honey production and pollination (approx. 1 h and 3 min)	Persuasion techniques View Bee Movie scenes and discuss • Effects of stopping honey production and pollination (approx. 1 h and 3 min) Devise scenes: "If bees could talk" At the relationship counsellor—humans and bees meet to discuss their issues
4	5Es—elaboration	Creative thinking: using research, generate multiple solutions for the problem drawing on scientific research Improvise and experiment with multiple ways of persuading the Diva tapping into appeals Critical thinking: select a solution and a persuasive technique for the final performance	
5	5Es—elaboration	Devise and rehearse performance	
6	5Es—evaluate	Performances De-enroll: Diva's wedding photos. Huge success!	

The combination of disciplines provides opportunity for interdisciplinarity and the richness provided by offering two ways of understanding and communicating knowledge; drama and science. In the Explore and Explain phase we recognise the need for deeper disciplinary knowledge. In order to transition students to genuine interdisciplinary understandings of the unit content, we need to ensure adequate disciplinary knowledge is fostered prior to synthesising the knowledge from both disciplines into a performance. Paige et al. (2008) assert the importance of introducing students to each discipline, highlighting their uniqueness, including discipline specific concepts, ways of thinking and working, and their skills and values. Equally students benefit from seeing how the disciplines "can complement each other when addressing complex issues" (Paige et al., 2008, p. 2).

Conclusion

This chapter offers a theoretical approach regarding the mutuality and inter-relativity of drama and science. Offered here is a way of conceptualising interdisciplinary practice for drama and science. We have demonstrated the mutualities of the two disciplines through the unit 'Inter-relationships in Our World'. By using a common context, each discipline's content is unified. Drawing together the values and approaches of both disciplines, offers a richer and more complete understanding. This approach embraces both discursive and non-discursive expression which allows the two disciplines exist in a symbiotic relationship, mutually benefitting each other.

Drama and science are brought together through project-based learning featuring the 5Es inquiry model and Mantle of the expert, culminating in a single assessment item, a multi-modal performance. The performance challenges students to demonstrate both discursive and non-discursive understandings of the importance of bees in ecosystems and how humans have impacted bees and as result the balance of the ecosystem.

We offer here a way of designing a unit of work that recognizes the value of both disciplines, neither discipline in service of the other. Drama and science are here positioned as mutual and enriching each other.

References

Abbs, P. (1987). *Living powers: The arts in education.* London & New York: Falmer Press.

Andersen, C. (2004). Learning in "as-if" worlds: cognition in drama in education. *Theory into Practice: Developmental Psychology: Implications for Teaching, 43*(4), 281–285. https://doi.org/10.1207/s15430421tip4304_6

Braund, M. (2015). Drama and learning science: An empty space? *British Educational Research Journal, 41*(1), 102–121.

Braund, M., & Reiss, M. J. (2019). The 'great divide': How the arts contribute to science and science education. *Canadian Journal of Science, Mathematics and Technology Education, 19*(3), 219–236. https://doi.org/10.1007/s42330-019-00057-7

Booth, E. (2012). Take-always from the world's first international teaching artist conference. Retrieved from https://teachingartistjournal.wordpress.com/

Bruner, J. S (1966). *Toward a theory of instruction.* Harvard University Press.

Bundy, P. (2005). Asking the right questions: Accessing children's experience of aesthetic engagement. *Applied Theatre Researcher, 6*, 1–12.

Bunterm, T., Lee, K., Ng, J., Srikoon, S., Vangpoomyai, P., Rattanavongsa, J., & Rachahoon, G. (2014). Do different levels of inquiry lead to different learning outcomes? A comparison between guided and structured inquiry. *International Journal of Science Education, 36*, 1937–1959. https://doi.org/10.1080/09500693.2014.886347

Bybee, R. W. (1997). *Achieving scientific literacy: From purposes to practices.* Heinemann.

Bybee, R. W. (2015). *The BSCS 5E instructional model: Creating teachable moments.* NSTA Press, National Science Teachers Association.

Coffman, T (2017). Inquiry-based learning: designing instruction to promote higher level thinking, 3rd edn. Rowman & Littlefield.

Crawford, B. A. (2014). From inquiry to scientific practices in the science classroom. In N. G. Lederman & S. K. Abell (Eds.), *Handbook of research on science education* (Vol. II, pp. 525–541). London, United Kingdom: Taylor & Francis Group.

Dewey, J. (1986). Experience and education. *The Educational Forum, 50*(3), 241–252.

Dorion, K. R. (2009). Science through drama: A multiple case exploration of the characteristics of drama activities used in secondary science lessons. *International Journal of Science Education, 31*(16), 2247–2270. https://doi.org/10.1080/09500690802271699

Ebbinghaus, H. (1913). *Memory: A contribution to experimental psychology* (H. A. Ruger & C. Bussenius, Trans.). New York, NY, US: Teachers College Press.

Ewing, R. (2011). *The arts and Australian education: Realising potential* (58). Retrieved from Camberwell, Victoria. https://research.acer.edu.au/aer/11/

Hammer, D. (1997). Discovery learning and discovery teaching. *Cognition and Instruction, 15*(4), 485–529. https://doi.org/10.1207/s1532690xci1504_2

Heathcote, D. (1995). *Drama for learning: Dorothy Heathcote's mantle of the expert approach to education*. Heinemann.

Heathcote, D., & Herbert, P. (1985). A drama of learning: Mantle of the expert. *Theory into Practice, 24*(3), 173–180.

Kidman, G., & Casinader, N. (2017). *Inquiry based teaching and learning across disciplines: A comparative theory and practice in schools*. Springer.

Langer, S. K. (1953). *Feeling and form: A theory of art*. Charles Scribner's Sons.

Langer, S. K. (1957). *Philosophy in a New Key: A study of the symbolism of reason, rite and art*. Harvard University Press.

Lerman, L. (2014). *Hiking the horizontal field notes from a choreographer*. Wesleyan University Press.

Lindvig, K., & Ulriksen, L. (2020). Different, difficult and local: A review of interdisciplinary teaching activities. *The Review of Higher Education, 43*(2), 697–725.

Loewenstein, G. (1994). The psychology of curiosity—A review and reinterpretation. *Psychological Bulletin, 116*(1), 75–98. https://doi.org/10.1037/0033-2909.116.1.75

Marx, R. W., Blumenfeld, P. C., Krajcik, J. S., & Soloway, E. (1997). Enacting project-based science. *The Elementary School Journal, 97*(4), 341–358. https://doi.org/10.1086/461870

Marzano, R. J. (2010). *Highly engaged classroom*. Marzano Research Laboratory.

Montouri, A. (2012). Creative inquiry. In N. M. Seel (Ed.), *Encyclopedia of the sciences of learning*. Boston, MA: Springer.

Murray, J. (2009). *Nondiscursive rhetoric: Image and affect in multimodal composition*. Retrieved from http://ebookcentral.proquest.com/lib/qut/detail.action?docID=3407478

Najami, N., Hugerat, M., Kahlil, K., & Hofstein, A. (2019). Effectiveness of teaching science by drama. *Creative Education, 10*, 97–110. https://doi.org/10.4236/ce.2019.101007

O'Connor, P. (2013). Drama as Critical Pedagogy: Re-imagining Terrorism. In M. Anderson & J. Dunn (Eds.), *How drama activates learning: Contemporary research and practice* (pp. 125–134). Bloomsbury Academic.

Ødegaard, M. (2003). Dramatic science. A critical review of drama in science education. *Studies in Science Education, 39*(1), 75–101. https://doi.org/10.1080/03057260308560196

Oliver, M., & Venville, G. (2012). Developing a 'thinking science classroom' . In G. Venville & V. Dawson (Eds.), *Art of teaching science: For middle and secondary school* (pp. 174–200). Allen & Unwin.

O'Neill, C. (1995). *Drama worlds: A framework for process drama*. Heinemann.

O'Toole, J., & O'Mara, J. (2007). Proteus, the giant at the door: Drama and theater in the curriculum. In L. Besler (Ed.), *International handbook of research in arts education* (pp. 203–218). Springer.

Paige, K., Lloyd, D., & Charters, M. (2008). Moving towards transdisciplinary: An ecological sustainable focus for science and mathematics pre-service education in the primary/middle years. *Asia-Pacific Journal of Teacher Education, 36*(1), 19–33. https://doi.org/10.1080/13598660701793350

Robinson, K., & Aronica, L. (2015). *Creative schools: Revolutionising education from the ground up*. Allen Lane.

Salomon, G. (1979). *Interaction of media, cognition, and learning/Gavriel Salomon*. Jossey-Bass Publishers.

Sciences, Engineering, & Medicine. (2005). *Facilitating interdisciplinary research*. Washington, DC: The National Academies Press.

Snow, C. P. (1959). *The Rede Lecture*. Cambridge University Press.

Sousa, D. A., & Pilecki, T. J. (2018). *From STEM to STEAM: brain-compatible strategies and lessons that integrate the arts*. Sage Publications.

Venville, G., Wallace, J., Rennie, L. J., & Malone, J. A. (2000). Bridging the boundaries of compartmentalised knowledge: Student learning in an integrated environment. *Research in Science and Technological Education, 18*(1), 23–35.

Verhoeff, R. P. (2017). The use of drama in socio-scientific inquiry-based learning. In K. Hahl, K. Juuti, J. Lampiselkä, A. Uitto, & J. Lavonen (Eds.), *Cognitive and affective aspects in science education research* (pp. 117–126). Springer.

Warner, C. D., & Andersen, C. (2004). "Snails are science": Creating context for science inquiry and writing through process drama. *Youth Theatre Journal, 18*(1), 68–86. https://doi.org/10.1080/08929092.2004.10012565

Willis, P. (1998). Notes on common culture: Towards a grounded aesthetics. *European Journal of Cultural Studies, 1*(2), 163–176. https://doi.org/10.1177/136754949800100201

Tricia Clark-Fookes (B.Ed., M.Ed., Ph.D.) is a lecturer based in the School of Creative Practice in QUT Creative Industries, Education and Social Justice Faculty. Her research and practice is situated in the field of Arts education with special interest in the practice and development of teaching artists. She is a member of the Drama study area but teaches across the range of arts curriculum areas. Tricia has trained as an artist and educator working in positions across the arts and education sectors. Working as a teaching artist, she continues to work with theatre companies both locally and nationally bridging the fields of arts practice and arts pedagogy.

Senka Henderson (B.Sc. (Hons), M.Sc., Grad. Dip. Ed., Ph.D., FHEA) is a lecturer in science education at the Queensland University of Technology (QUT), Brisbane, and teaches in undergraduate and postgraduate courses. Senka's first career included working in pharmacology, biochemistry and drug discovery, and she has worked in research laboratories investigating new discoveries in carbohydrate chemistry. In her second career as a science educator, she has worked on three ARC science education research projects exploring the emotions of pre-service science teachers in university settings and students in high school science classrooms. She has been involved in other projects related to context-based science education and mindfulness. She is the reviewer and assistant editor of the international journal Research in Science Education.

The Science Drama Project: Meaning in the Middle

Kitty van Cuylenburg

Abstract Under the theoretical framing of experiential education (Dewey, 1910) and constructivist learning (Vygotsky, 1978), this research seeks to explore the intersection of Science and Drama, to consider how interdisciplinary teacher practice might develop a more meaningful student experience and understanding of Science as a Human Endeavour (VCAA, 2016) for middle school students in Victoria, Australia. This research uses Self-study Methodology to situate myself—a teacher in my practice—in exploring the research question through self-reflection, dialogue and critical analysis. Data collection methods include using interviews, focus groups and a reflection journal, and engaging with Applied Thematic Analysis to find connections through the collected data. As a teacher-researcher I reframed my understanding of the problem: finding meaningful learning for students was not a result of interdisciplinary practice. I must first consider a pedagogical foundation for student voice, confidence, and authentic connection. I must enable students to ask a human question, and then revisit the experience with students at a later date. Meaning was found not in the moments of the program, but in the meditation on it, and illuminated the critical nature of supporting students to reflect on their learning and experiences.

Keywords Science drama · Interdisciplinarity · Pedagogy · Self-study methodology

Introduction

Teaching and Learning Context

Middle secondary school for 13–15 year-olds is an interesting place. Developmentally, adolescents are finding their emerging adult selves, negotiating those selves amongst each other, and grappling with many questions and shifting identities. Working in such an environment is emotional for both teachers and students, as

---

K. van Cuylenburg (✉)
School of Education, Deakin University, Burwood, VIC, Australia
e-mail: k.vancuylenburg@deakin.edu.au

P. J White et al. (eds.), *Science and Drama: Contemporary and Creative Approaches to Teaching and Learning*, https://doi.org/10.1007/978-3-030-84401-1_3

teachers' try to build introspection amongst students' complex inner lives (cognitively, hormonally) and complex outer lives. I entered this environment for the first time as an associate teacher (still in training) with a background in both science and theatre, experience in an ed-tech start-up, and as a mother to 2 year-old girl. I brought multiple selves into my classroom—the analytical and iterative scientist, the curious and emotional artist, the innovative and forward-focused entrepreneur and the patient and practical mother. These selves collided within my first term of teaching in 2016 as I negotiated my first experience of teaching Year 8 Drama.

I taught students who attended a low-SES, multicultural, outer-metropolitan school in Melbourne, Victoria. One of my drama classes was the 8A class, which was an accelerated curriculum and enrichment class. Throughout Term 1 I became aware of the need to extend students in this class beyond the school curriculum—recognising Black's (2006) research into how disadvantaged schools which have high student achievement do so because they have a student-centred approach to teaching and learning. I wanted to create a program of learning which would centre the students and broaden their experience of learning in the classroom. So, lacking teaching experience I drew on the resources I had at the time—my understanding of how to create an ensemble theatre piece, my content familiarity with science, my passion for combining arts and sciences, and my entrepreneurial streak to try something new regardless of 'failure', thus the Science Drama project was created.

The Science Drama Project

Science Drama is a process whereby students explore scientific concepts through dramatic practice (Yoon, 2006) which may or may not result in a performance of their product. In both 2016 and 2017, I developed a Science Drama project for the 8A class across Term 2, culminating in a performance at the annual Winter Concert. In each year the participating class generated their own scientific inquiry question based on their interests, and in response to this created a dramatic work exploring the results of their investigation. In 2016 this project investigated 'What happens after you die?' and in 2017 'Why do people act the way they do?'

To enact the program of learning, I collaborated with the 8A science teacher so that each week either one of the 8A drama or science lessons would be devoted to this project. This collaboration developed through our shared passion for interdisciplinary learning in science, and our joint need to fulfill our requirements to extend the 8A class in both their drama learning and science learning. We would each focus our sessions on our subject specific disciplines—either investigating the science of the research question or extending students to explore the concept through embodied work. Towards the end of the project we would work together in lessons as we neared performance. Qualified as both a drama and science teacher, I managed the project and integration of the learning between the disciplines. Table 1 presents the framework for the program of learning.

Table 1 Science Drama Project Learning outline

Week	Drama	Science	Pedagogy
#1	Lesson 1: What is science drama? What question should we ask?		Process drama
#2		Lesson 2: Whole class exploration of the science behind the question.	Inquiry based learning
#3	Lesson 3: Exploring the humanity in the science 'Science is a human endeavour'.		Improvisations Performance structures
#4		Lesson 4: Small group research on chosen key parts of science.	Research skills
#5	Lesson 5: Developing scripts for each act/scene, focus on showing, not telling.		Non-naturalism Playwriting skills
#6		Lesson 6: Developing science in the scripts.	Scientific literacy/vocabulary
#7	Lesson 7: Rehearsals. Students using tablets to film/revise work.		Filming for feedback
#8/9/10	Lesson 8–10: Rehearsals and Peer Feedback 1. So Fun: What was funny or entertaining? 2. Review: What needs to be improved in this scene? Why? 3. You See: What scientific information was made clear? 4. Some More: Additional teacher feedback.		Voice-work Student feedback routines Performance skills Costuming and sets (projections/digital)
#11–12	Dress Rehearsal Performance: Winter Concert		

Pedagogical Context

In my context, Science Drama developed as an inquiry into a question through scientific and dramatic pedagogy. The beauty in the combination is the ability of the educator(s) to employ the specific pedagogies which best enhance the drama and science, and which work contextually for their class. Evidence of this is how science and drama connect differently in every classroom—from improvisations and still images in life sciences (Taneri & Engin-Demir, 2013), prose and scripts in the geology classroom (Keech, 2001), Readers Theatre, role-play and debate (Yoon, 2006), teacher in-role in the environmental science classroom (Stevenson, 2014), script-writing and play-building in junior school physics (Nicholas & Ng, 2008) to dramatic response to lab investigations (Hegedus et al., 2016). These different approaches lead to different shifts in learning—the use of dramatic pedagogy to explore scientific questions has made visible invisible scientific concepts such as gravity (Bort, 2007; Braund, 2015; Taskin-Can, 2013), or supported students to explore yet to be experienced socio-scientific phenomena such as the ethics of genetic testing (Dorion, 2009). The use of 'Mantle of the Expert' has enhanced student inquiry (Fraser et al., 2012), freeze frames and role-play has given clearer understanding of previously inaccessible scientific concepts (McGregor, 2014), and performance has boosted engagement with environmental science issues (Curtis et al., 2013)—which is part of my rationale for choosing to include performance in my project. The denominator underlying each of these studies is the desire for improved student learning; of importance is how teachers can heed this research to deliver student-centred, applied and inspiring learning in their classes.

Recently, research into the coaction of drama and science pedagogies has been through mixed method research (Curtis et al., 2013; McGregor, 2014), and case studies (Dorion, 2009; Fraser et al., 2012). Both of these methodologies are conducted from an outsider's research perspective—research is missing which builds on the benefits of combining science and drama through exploring the experiences of the teacher-researcher and students situated in it. Grounded in the learning theory of experiential education (Dewey, 1938), I engaged in the Science Drama project not in the 'traditional' sense of students acquiring knowledge, but 'progressively' in students generating knowledge through facilitated experiences. Through using self-study methodology, I intended a deeper exploration of the in-context connections between teacher practice, student experience and interdisciplinary thinking through the question:

> To what extent does interdisciplinary teacher practice in science and drama enhance student understanding of science as a human endeavour and facilitate meaningful experience in a middle-school classroom?

Both Ways Thinking and Making Meaning

In 1910, Dewey developed his seminal work on 'How We Think', his thesis of learning based in experimental inquiry stemming from curiosity. This was progressed by Vygotsky (1978) who held a premise that learning occurs regardless of the application of school and is a natural outflow of growth and development. Science Drama is an attempt to provide space in school for the natural inquiry of adolescents as part of this growth and development—where learners come to understanding through the interactions of immediate phenomena, historical experience and presented ideas in the constructivist tradition (Richardson, 1997).

Pedagogically, this has opened up the use of drama to explore student experiences and scientific ideas, and as a way to overcome 'transmission' modes of teaching—where information is presented without exploration of prior and current knowledge (Richardson, 1997). For this project I viewed dramatic pedagogy as a way-finding tool to create meaningful dialogue about phenomena to guide student inquiry; and as a process for allowing the brain multi-sensory inputs to integrate knowledge and progress education (Adey & Dillon, 2012). I also used dramatic pedagogies to decentralise the teacher in the class in order to more fully realise constructivist student engaged learning (Taskin-Can, 2013), and in the science classes, students engaged in pedagogy which used social dynamics (small peer groups) to engage construction of knowledge (McGregor, 2014).

Dramatic pedagogies are also highly social, meeting adolescents at their point of need. Drama has long been seen as a means of personal development (Way, 1998), a way for students to lead their own inquiry via play making (O'Neill, 2015), and as a pedagogical technique for students to learn other content (Bolton, 1979). Exploring how teacher practice of dramatic pedagogies can enhance science understanding has developed as an evolution of Bolton's (1979) research in recent decades (Dorion, 2009; Lee et al., 2015). Lampert (2006), Chang et al. (2015), Leahy (2016), and Uzuncz and Demirhan (2017) have all demonstrated the progression of critical thinking through using arts or drama-based pedagogies, and the importance of intertwining dramatic-creative-divergent and scientific-critical-convergent approaches to provide a wholistic toolbox for problem-solving and ideas generation. Thus, Science and Drama can be seen through the same pedagogical lens: students' natural development leads inquiry in both the social and scientific (Vygotsky, 1978). It supports Everett's (2016) summary of interdisciplinary education as one which develops the whole student and focusses not just on the intellectual, but also on the personal and social; and intertwining these foci with a decentralised teaching model can generate understanding in both these subjects via constructivist principles.

However, current research does not address how the combination can generate meaning for learners. The closest is by Snyder (1990), who undertook a longitudinal study into the thinking patterns of 50 students at Massachusetts Institute of Technology in the US, specifically exploring the balance between two distinct modes of thought: 'Mode One', finding truth through hard data where outcomes are certain; and 'Mode Two', outcomes are relative, truth can be revised, and it is through empathy and intuition that people come to 'knowing'. Snyder (1990) found that those Mode One undergraduates who learned to shift their thinking across the years to using Mode Two, found deeper meaning in work and relationships. This research suggests that as educators, if we can support students to develop 'both ways thinking' we are building up student capacity for future meaning and satisfaction in life. The Science Drama project is a start towards this, facilitating a balance of Mode One (scientific/cognitive thinking) and Mode Two (dramatic/emotional-social thinking). I wanted to know how the combination, enacted by interdisciplinary teacher pedagogy (if at all) creates a meaningful student experience and deeper connection to science as a human endeavour. I wanted to find meaning for middle school students and meaning within interdisciplinary learning.

Methodology

An observed gap in Science Drama research is how interdisciplinary teacher pedagogy, specifically in an Australian context, can be more deeply understood through problematising teacher practice and processes. Hannigan et al. (2016) showed the promise of self-study to explore interdisciplinary work, Coleman and Lieder (2014) demonstrated its applicability to truly deepen teacher practice, and Pinnegar and Hamilton (2009) showed it assists teacher researchers to understand their practice within their teaching context. For my research, the application of self-study methodology was to expose the problems in my practice and critically evaluate how I could link the disciplines to create a meaningful experience for my students. The self-study methodology used was based on the Five Foci from Samaras (2011): author your own situation question, collaborate with critical friends, plan new pedagogies for improved learning, make transparent the research process, and present generated knowledge. The intention in this methodology is for me, the teacher, to reframe my own understanding rather than evaluate the impact of the research (Lassonde et al., 2009).

Data collection was through two collegial interviews (with the collaborating 8A science teacher, and an interview with another science teacher at the school) and student focus groups (a focus group of four students from the 2016 cohort, a focus group of 6 students from the 2017 cohort,) and a reflective journal from across Term 2, 2017. Collection of data occurred through digitally recording the discussions—I posed questions in a semi-structured interview format (i.e. Somekh & Lewin, 2012). These conversations were later transcribed into written format and de-identified. The research journal is a collection of reflections I wrote on the Science Drama program throughout the 2017 unit of work, specifically responding to the questions also posed to the participants. Ethics was obtained through Deakin University, Australia, with appropriate permissions granted by the study school.

From there, inductive analysis was used to make sense of the data—to find patterns and relationships from and between observations to generate a tentative hypothesis in conclusion. Data was sorted via emergent themes, and then iterations of understanding were developed as I navigated the data (Samaras, 2011), continually realigning my thoughts to the research question (Stake, 2010). This research approached analysis in the way of Guest et al. (2012) who developed an approach called Applied Thematic Analysis. Its focus is on exploring a practical problem and employing whichever analysis methods might be useful to "get the analytic job done in a transparent, efficient, and ethical manner" (Guest et al., 2012, p. 18).

Findings and Discussion

In an unexpected blow to my ego, it became apparent that for my students, the important parts of the Science Drama project were not actually about Science or Drama. There was an enormous amount of data which pointed to students valuing the process, not the disciplines of knowledge. Three themes were strong here: that students valued ownership and agency over the project, that they sought authentic connections with each other and their teachers, and that they were primed to seek praise and boost their confidence through the project. Students were curious to connect to each other wherever they could, and this was also reflected in the questions they explored. Rather than using the Science Drama project to explore a scientific concept like magnetism or atomic theory, the 2016 cohort asked 'What happens after you die?' and the 2017 cohort asked 'Why do people act the way they do?' The inclination of students was to use the project to connect to their humanity, and both teachers interviewed saw that unfolding. Teacher 2 reflected that: *"Five days a week they spend at school and school is a very social place. And I think that ... kind of makes them curious like 'Oh why did that person do that?'... they genuinely want to know"*. *"I think it makes sense like they're trying to make science real for them. It is kind of acting as an enabler to go 'How does science fit into my everyday life?'"* suggested Teacher 1.

Students articulated that they also thought that the connection between their science questions and their lives was key: *"Because a lot of stuff [at school] a lot of us think that it's not useful when we're older. And I feel like if somehow it could be displayed to use that it's going to be useful like a lot of people would try more"* (Y8Student3) and *"[We saw] how it affects a person rather than just learning that in science class"* (Y9Student4). What I observed in the Science Drama projects were that students were trying to create a narrative to explore how they felt about the world around them, and they were using the cognitive concepts in science to navigate how they felt about a lived phenomenon: *"You learned to accept it ... after seeing the process ... you connect to it"* (Y8Student4). Teacher 1 said that students were *"Using science to tell a story"*, which in turn comes back on itself as through stories, science emerges as a human endeavour and students are offered insight (Ødegaard, 2003).

Ownership, Agency and Connection

Ødegaard's (2003) synthesis also demonstrated that when students are charged with their own design and presentation of a concept, they think about it in ways which are meaningful for them in that moment and thus take ownership of the idea. Both the 2016 and 2017 cohorts were upfront in their disengagement with their idea if they felt the direction of inquiry was not set by them. Students identified with their question choice being an important part of the project—Y8Student4 was adamant that the most important thing was *"Don't pick a stupid question!"* In addition, students elaborated on their desire to have ownership of more decisions in the process: their inquiry groups, choice of materials, lesson content, format of presentation, and the number and connection of lessons in the program. Where students felt like they had those moments of ownership, they felt that *"You feel free to do it. You feel like it's your own"* (Y8Student2).

I took a structured approach to the Science Drama sequence, planning the framework but allowing students flexibility within that to keep the program moving forward. When I allowed the 2017 cohort of students to self-manage *"Groups struggled to find the words and were distracted throughout, and student groups imploded due to tensions in the class"* (Journal2505), and students admitted that: *"If you have a lot of freedom you get off topic, you don't always eventually do the work"* (Y8Student1). There was also a delicate balance between my intervention in student groups to support effectiveness, and the agency of students to own the process and manage their groups themselves. *"Some people didn't engage and participate…because they didn't like the people in their group"* (Y9Student4). In both cohorts, students recognised the tension between wanting to work with their friends, but the realities of that impacting on their work.

The need for, and the nature and intensity of intervention thus depended on the particular student mix. Unlike the 2016 cohort, for the 2017 cohort the need for intervention was significant. The nature of the ensemble process and the question ('Why do people act the way they do?') exposed existing issues (i.e. racism, oppression) the class was grappling with. Their science teacher said that the Science Drama process: *"Gave that space for them to have more of a voice in the class, more of an opinion, more of a space to discuss, to think" (Teacher2).* It was in this environment that these issues could erupt, which resulted in the whole class requiring a 'circle time' session with the school counselling team *"for the whole of the next lesson to reconcile student behaviours [amongst one another]"* (Journal2505). Further tension was added to the project as the outcome, a public performance, was decided for the students. Some students thought performance was important, but many others thought it hobbled their development: *"I think because we have that knowledge of being in front of an audience we sort of, you know, tried to back out as much as we could so we didn't have to do it."* (Y9Student3).

In each year, there was a moment where I wondered if we should keep persisting given these tensions. Both years we did, and I am glad to have done so both for student development and my own learning. In our reflections, both the science teacher and I thought that the students' commitment and performance skills lifted significantly in the performance over the rehearsals: *"I think generally like 95% of kids were on board because it was kind of like we're learning this for someone else, to show someone else"* (Teacher2). I realised that a performance outcome may be a valid approach, but only when students feel like it is their choice, and otherwise I should see the development of my students' inner worlds as the outcome (Montuori, 2006). I was so focused on doing Science Drama, I missed how the process translated into students' being: *"[I know now] not everything is about you"* (Y8Student2) and *"With the way you did it I think we broadened our like understanding of other people"* (Y9Student3).

New Understandings

The Year 9 Cohort were naturally more introspective than the Year 8 Cohort during data collection. They had time away from the process (unlike the Year 8 Cohort who had just completed it) and were able to articulate the insight the experience had given them. A small focus group environment provided that safe space which was not present in the classroom at the time and allowed students to put forward feedback concerning their needs to me as their teacher. Both the 2016 and 2017 cohort were focused on their feelings, but through different lenses. The 2017 cohort was focused on the way the project would have felt if they had had more ways to build their confidence, and also give and receive feedback in a safe way. *"Maybe in Drama we could have one or two lessons where we focus on confidence building and like learning to gain confidence"* (Y8Student3). In the 2016 cohort however, one student found that it was actually the combination of Science and Drama which was the tool needed to access their confidence. *"[The Science Drama process] gets you out of your comfort zone to express who you are and what you value and stuff. You can bring your personality into science, and I think it made us more confident"* (Y9Student3). It might be for some of my students that the combination of Science and Drama was a catalyst for confidence, but to better support all students I needed to intentionally structure pedagogy for confidence building, and also develop safe feedback processes for students to hear it from their peers. In turn, this would also build students' belief in their own potential for success in Science Drama practice (Nicholas & Ng, 2008).

What stunned me the most was the broader reflections of the 2016 cohort. They had opened themselves up to the process in ways I had not anticipated. The Science Drama process enabled them to grapple with their experiences, both past and future, and empathise and connect with others. *"It's given me a better understanding [of death] and I'm not worried"* (Y9Student1). *"For some people having deceased grandparents, or [not] having their younger brothers or sisters when their parents had miscarried or something ... find a way to understand what could have happened...and where they could be now"* (Y9Student3). *"It helps me appreciate those people more ... I appreciate the nurses for trying to do their job, appreciate detectives trying to figure out stuff like that. Helps me appreciate what their choice was in life to do... Be able to understand their circumstances as well"* (Y9Student2).

Teacher 2 aptly commented that *"When a kid has to have a conversation about something, that's when they truly understand it"*, and I saw these students work through their understanding in our conversation. They appreciated the time to reflect on it, *"Other than now, we wouldn't have discussed it"* (Y9Student4), and thought that the 2017 cohort should also receive the time and space to do so a year after the project. The importance of time plus the conversational format was illustrated in the lack of connective reflection in the 2017 cohort. Towards the middle of the process I asked the 2017 students to complete a written reflection on the Science Drama process, so I could see where we were and how students were feeling. This yielded minimal insight. It appears that time, plus the space to work through their thinking processes unencumbered by translating them into writing, developed the capacity for students to make those meaningful connections.

Combining Drama and Science

It was interesting trying to understand how students and teachers perceived the project through each subject, and whether or not combining two previously uncombined disciplines (at my school) yielded greater insight or meaning in student learning. Students said that science thinking is "*Interesting*" (Y8Student6), "*Observ[ation]*" (Y8Student2), "*Stressful*" (Y8S1), and "*Questioning*" (Y9S3), and Drama thinking is "*Playful*" (Y8Student2), "*Confidence*" (Y8Student4), "*Quick*" (Y8Student2), and involves teamwork and discussion (Y9Student1). It appeared that these approaches slowly settled over time into a truce, ebbing and flowing along the journey. *"I think it has those two disciplines kind of co-existing and students jumping between the two"* (Teacher1). *"It's four fifths drama and a fifth science"* (Y8Student3). *"At the start it's research, generating, finding—the drama part was when we have to act and plan—it was both"* (Y9Student2). *"It might start with a drama and then the students go explore the science ... I think it's probably a spectrum"* (Teacher1). *"We're learning. Seriously though ... like really it is that they're learning ... I can't say that it's science or drama. It's both. It's everything it's just students, kids learning" (Teacher2).*

I assumed that the shared excitement or novelty of getting involved in a different kind of process would foster 'buy-in' from students, a reflection of my own history and the work of Dorion (2009) and Fraser et al. (2012) which demonstrates that interdisciplinary learning with drama and science promoted 'buy-in' from students. In my context, this did not overcome the inertia of Science Drama as: *"Another thing we have to do at school"* (Y8Student4). It took doing the whole process and hindsight before some students were comfortable saying that: *"I actually found it really exciting"* (Y9Student4), or *"I think it was fun towards the end you know"*. (Y9Student2). Or even: *"Basically it was really fun because you do a lot of moving and not much thinking [in drama] in my opinion. But then in science you do more writing and thinking which is basically the opposite of drama"* (Y8Student2). *"So technically when you combine them it should be like a perfect combination"* (Y8Student3).

Evidence from data showed that the science and drama combination worked for these students, but their main focus was not on particular pedagogy, the particulars of science inquiry or dramatic practice. Students asked a 'human question' and then focused on how they could have ownership of the project, how they could connect with each other and have opportunities to reflect on their learning and build confidence in their abilities. It became clear that to students, the importance of interdisciplinary pedagogy in science and drama is secondary to learning ways to learn with and about each other.

Meaning in the Middle

The Science Drama project was structured to build interdisciplinary thinking across drama and science; however, the purpose was not for greater learning or skill attainment (i.e. Dorion, 2009; McGregor, 2014; Nicholas & Ng, 2008) but to generate meaningful experience and understanding for students. What I found was that the underlying benefits of the Science Drama *process* were that authentic connections were fostered and inquiry learning felt more genuine. The benefits of the science and drama *combination* were that it let students ask social-emotional questions under the security of scientific inquiry; and it enabled teachers and students to recognise and appreciate each other's capabilities in the broad skillset present across science and drama. But when it comes to generating meaning, I found that in my context this was an evolution. First the student finds a place within the process (ownership, connection, confidence) assisted by a considered and reflective teacher practice. Secondly, students need to engage with the inquiry and be willing to discuss their ideas, connect those ideas with their experiences and develop their creative response. Lastly, given space to let the process settle and time to reflect on the experience, some students might find that deeper meaning emerges.

Smagorinsky (1996) states there are two ways in which students change during creative pursuits, firstly the change which occurs during the process itself, but secondly the rumination, the reflection on the experience. Ødegaard's (2003) position was that the most valuable role a teacher plays in the Science Drama process is in guiding the students through reflection after the fact on how their experience connects with their own life. Ostern and Kristofferson (2015) considered the combination of artistic expression, storytelling and science to have the potential to deepen meaning as students remembered their experiences and connected them to the universe. It appears that in conducting this research I uncovered the reality of reflection—the 2016 focus group showed that given time away from the project, the creation of a safe space, and the ability to reflect through a dialogue, students were able to construct meaning from the experience. Unlike Ostern and Kristofferson's (2015) work, this dialogic approach was able to capture those moments of meaning where other data collection methods may not.

Conclusion

This research does not try to investigate the effect size of interventions or amount of meaning that the Science Drama project generated for students; that is not its purpose. The intention of my self-study is to reframe my understanding of the problem rather than its impact (Lassone, 2009) and how I needed to alter my framework and practice to better serve the students I teach (Pinnegar & Hamilton, 2009). It was an emotional process, as in the examination I exposed my own biases, ego and motivations for Science Drama and developed a new appreciation for the breadth and depth of reflective practice that teachers need engage with to serve the students in their context. My scientist self was trying to ‘solve’ the problem that there was a particular pedagogical practice in Science Drama which would promote student understanding and meaning making; which was fuelled by my biases as a teacher and an enthusiast of both disciplines. I had to see my voice as only one of 12, and listen to the criticisms and find through lines of similar data across all 12 participant voices so that my findings were authenticated and my biases were tempered (LaBoskey, 2004). I now understand why some researchers might not undertake self-study methodology—as White (2017) stated, without the willingness to publicly and vulnerably analyse their practice, actions and teacher identity, self-study is unlikely to yield significant improvements either for the researcher or the wider research community. To take this research forward, I would seek to understand how this reflection and personal change might influence the project for my next cohort of Science Drama students or explore this problem via a collective of teacher researchers involved in Science Drama, through the process of collective self-study.

Now, I no longer frame meaningful learning for students in Science Drama as a localised problem of bringing together the right combination of pedagogy, student experiences and thinking processes in class. Pedagogy is important, but for my students, finding meaning in Science Drama was about playing the long game. I saw that meaningful learning in Science Drama needs to begin long before the creative inquiry starts, developing a robust classroom pedagogy and engaging student voice, confidence and authentic connections with and between teachers and students. Meaningful learning continues by asking a human question, where students can discuss their varied, deep experiences of life in a safe space and engage with dramatic processes. This research illuminated that in the Science Drama projects, students were thinking about their loved ones who had passed away, their baby siblings they had never met, their underlying beliefs about racism and the school system which they see oppresses them. They were connecting science into their lives, large ideas which percolated long after the creative inquiry finished.

Lastly, the smallest part of the process reveals the greatest insight, and importantly, did not happen during the program or as a result of interdisciplinary pedagogy. This study showed that students required time and space to let the Science Drama project settle, to then come back to reflect on the experience and their new understandings. Had I not used focus groups for data collection, I may not have experienced this reflection in action. And this is the long game—connecting with your students long after learning and allowing them the space to process their thinking, uncover their understanding, and excavate their meaning in the middle of the experience.

References

Adey, P., & Dillon, J. (Eds.). (2012). *Bad education: Debunking myths in education*. McGraw-Hill.

Black, R. (2006). *Crossing the bridge: Overcoming entrenched disadvantage through student centred learning*. Education Foundation Australia.

Bolton, G. M. (1979). *Towards a theory of drama in education*. Longman.
Bort, N. (2007). Using the arts to enhance science learning. *Science Scope, 31*(1), 56.
Braund, M. (2015). Drama and learning science: An empty space? *British Education Research Journal, 41*(1), 102–121.
Chang, Y., Li, B., Chen, H., & Chiu, F. (2015). Investigating the synergy of critical thinking and creative thinking in the course of integration activities in Taiwan. *Educational Psychology, 35*(3), 341–360.
Coleman, E., & Leider, M. (2014). Personal and professional growth realised: A self-study of curriculum design and implementation in a secondary science classroom. *Studying Teacher Education, 10*(1), 53–69.
Curtis, D., Howden, M., Curtis, F., McColm, I., Scrine, J., Blomeld, T., Reeve, I., & Ryan, T. (2013). Drama and environment: Joining forces to engage children and young people in environmental education. *Australian Journal of Environmental Education, 29*(2), 182–201.
Dewey, J. (1910). *How we think*. D.C. Heath & Co.
Dewey, J. (1938). *Experience & education*. Kappa Delta Pi.
Dorion, K. (2009). Science through drama: A multiple case exploration of the characteristics of drama activities used in secondary science lessons. *International Journal of Science Education, 31*(16), 2247–2270.
Everett, M. (2016). Fostering interdisciplinary understanding and skills. *Currents in Teaching & Learning, 7*(2), 28–37.
Fraser, D., Aitken, V., Price, G., & Whyte, B. (2012). Inquiry learning, drama and curriculum integration. *Teaching and Learning, 3*, 32–40.
Guest, G., MacQueen, K., & Namey, E. (2012). *Applied thematic analysis*. Thousand Oaks, CA: SAGE.
Hannigan, S., Raphael, J., White, P., Bragg, L., & Clark, J. (2016). Collaborative reflective experience and practice in education explored through self-study and arts-based research. *Creative Approaches to Research, 9*(1), 84–110.
Hegedus, T., Segarra, V., Allen, T., Wilson, H., Garr, C., & Budzinski, C. (2016). The art-science connection: Students create art inspired by extracurricular lab investigations. *Science Teacher, 83*(7), 25–31.
Keech, A. (2001). *Making a play for science*. Retrieved from www.TeachingK-8.com
LaBoskey, V. K. (2004). The methodology of self-study and its theoretical underpinnings. In J. J. Loughran, M. L. Hamilton, V. K. LaBoskey, & T. L. Russell (Eds.), *International handbook of self-study of teaching and teacher education practices* (Vol. 12, pp. 817–869). Springer.
Lampert, N. (2006). Critical thinking dispositions as an outcome of art education. *Studies in Art Education, 47*(3), 215–228.
Lassonde, C., Galman, S., & Kosnik, C. (Eds.). (2009). *Self-study research methodologies for teacher educators*. Sense Publishers.
Leahy, K. (2016). Winning the future: An investigation into the creative capacity across the levels of education in Ireland. *Creativity Research Journal, 28*(2), 188–197.
Lee, B., Patall, E., Caution, S., & Steingut, R. (2015). The effect of drama-based pedagogy on PreK-16 outcomes: A meta-analysis of research from 1985–2012. *Review of Educational Research, 85*(1), 3–49.
McGregor, D. (2014). Chronicling innovative learning in primary classrooms: Conceptualising a theatrical pedagogy to successfully engage young children learning science. *Pedagogies: An International Journal, 9*(3), 216–232.
Montuori, A. (2006). The quest for a new education: From oppositional identities to creative inquiry. *ReVision, 28*(3).
Nicholas, H., & Ng, W. (2008). Blending creativity, science and drama. *Gifted and Talented International, 23*(1), 51–60.
Ødegaard, M. (2003). Dramatic science: A critical review of drama in science education. *Studies in Science Education, 39*(1), 75–101.
O'Neill, C. (2015). *Dorothy Heathcote on education and drama—Essential writings*. Routledge.

Ostern, A., & Kristofferson, A. (2015). Combining art and science in exploration of humanity and the universe: Teenagers' storied experience of the universe played back in improvisations theatre in a learning context. *Applied Theatre Research, 3*(3), 251–270.

Pinnegar, S., & Hamilton, M. L. (2009). *Self-study of practice as a genre of qualitative research: Theory, methodology, and practice*. Springer.

Richardson, V. (1997). *Constructivist teacher education: Building new understandings*. Falmer Press.

Samaras, A. (2011). *Self-Study teacher research: Improving your practice through collaborative inquiry*. SAGE Publications.

Smagorinsky, P. (1996). Multiple intelligences, multiple means of composing: An alternative way of thinking about learning. *NASSP Bulletin, 80*(583), 11–17.

Snyder, B. (1990). Literacy and numeracy: Two ways of knowing. *Daedalus, 119*(2), 233–256.

Somekh, B., & Lewin, C. (2012). *Theory and methods in social research* (2nd edn.). SAGE.

Stake, R. (2010). *Qualitative research: Studying how things work*. Guildford Press.

Stevenson, L. (2014). Lab coats, test tubes and animated expressions: Drama in a middle school science classroom. *NJ—the Drama Australia Journal, 38*, 64–73.

Taneri, P., & Engin-Demir, C. (2013). Implementation of life sciences from the perspectives of students: Creative drama as a qualitative data collection method. *Elementary Education Online, 12*(1), 267–282.

Taskin-Can, B. (2013). The effects of using creative drama in science education on students' achievements and scientific process skills. *Elementary Education Online, 12*(1), 120–131.

Uzunoz, F., & Demirhan, G. (2017). The effect of creative drama on critical thinking in preservice physical education teachers. *Thinking Skills and Creativity, 24*, 164–174.

VCAA. (2016). *Victorian curriculum F-10*. Victorian curriculum and assessment authority. Retrieved from http://victoriancurriculum.vcaa.vic.edu.au

Vygotsky, L. (1978). Interaction between learning and development. *Readings on the Development of Children, 23*(3), 34–41.

Way, B. (1998). Development through drama. Atlantic Highlands, NJ: Humanities Press (Originally printed 1967).

White, P. (2017, March 07). Limitations to self-study. Personal Communication.

Yoon, H. (2006). *The nature of science drama in science education*. Paper presented at the 9th international conference on public communication of science and technology. Retrieved from http://citeseerx.ist.psu.edu/viewdoc/download?doi=10.1.1.569.5065&rep=rep1&type=pdf

Kitty van Cuylenburg (B.A./B.Sc. (Hons), M.Teach.) is a VIT registered drama and science teacher, a Teaching and Leadership advisor for Teach for Australia, and theatre maker and board member of In The Park Productions. She has a background in mine-site hydrogeology, has worked for an education startup in NYC, performed in and produced theatre in Melbourne (MICF, MIFF), and has a passion for interdisciplinary practice of science and drama. She is an erstwhile research student at Deakin University, and undertakes action research in this space through her teaching practice.

The Treatment of Dr. Lister: Investigating and Revisiting a Famous Primary Science-Based Drama

Susan Davis

Abstract With ongoing calls to make science learning more engaging for students while building teacher confidence and capacity to teach it, there is value to be found in working with drama-based inquiry models to frame and activate science learning. The Dr. Lister drama provides an historical model that was documented by renowned drama educators Dorothy Heathcote and John Carroll and other drama educators in the 1980s. An analysis of that work provides the basis for this chapter and may serve to expand contemporary approaches to science education and learning. The use of drama and other artforms provides creative ways to engage children with science learning. Such a transdisciplinary approach also builds student understanding of the multiple roles of artist, scientist and entrepreneur which are integral to the success of any innovation. We can learn from these models of practice to engage in new investigations and inquiries and encourage students to be the creative problem-solvers needed for the current era, when the threat of pandemic and microorganism-based challenges once again have risen to the fore.

Keywords Drama pedagogy · Science history · Dramatic framing · Teacher-in-role

Introduction: Learning from Historical Models of Practice

Working with drama pedagogy in science education is not a new undertaking. In drama education circles there is a rather 'famous' primary or elementary school drama called the Joseph Lister drama and it was originally explored with primary school students in England the 1980s. It was developed by a legendary teacher named Dorothy Heathcote and one of her post-graduate students, then collaborator and friend, Australian drama educator John Carroll (Bolton, 2003; Davis, 2013). The project used drama and creative pedagogies to explore the significance of work such as that of Joseph Lister in microbiology and surgery and to also examine the difficulties innovators must overcome to bring about major changes in a field.

S. Davis (✉)
Central Queensland University, Rockhampton, Australia
e-mail: s.davis@cqu.edu.au

P. J White et al. (eds.), *Science and Drama: Contemporary and Creative Approaches to Teaching and Learning*, https://doi.org/10.1007/978-3-030-84401-1_4

An analysis and re-imagining of the Dr. Lister drama can provide a focus case and way in for exploring the possibilities for STEAM education in the 21st century. Through examining an effective pioneering program, contemporary educators can explore other affordances now possible in the context of contemporary developments, and through the various media and digital technologies. It also provides a means into examining contemporary issues and future challenges that exist in the field of medical sciences, including antibiotic resistance, pandemics and the spread of disease. The threat of global pandemic spread through microorganisms (in this case the COVID-19 virus) has become a grim reality. Student understanding of different microorganism, transmission and prevention, and the impact of simple actions such as handwashing are more relevant today than ever.

The question explored is:

How might we learn from the Dr. Lister drama to design contemporary learning, programs explore medical/scientific developments and engage students as active learners and inquirers?

For a contemporary primary school science educator, engaging with this historical work could be framed through science content knowledge in strands such as 'Science as a Human Endeavour' and 'Science Inquiry Skills' (ACARA, 2016) and the Biological Sciences. The historical work this chapter examines was conducted with primary school aged children in lower to middle primary school years. It is noted however that in the current Australian curriculum, for example, microorganisms are not addressed until year 9.

The work discussed in this chapter draws from the field of education and pedagogy known as drama in education. In terms of research framing, the work is a case study (Merriam, 1998; Yin, 2009) investigating a "a unit around which there are boundaries" (Merriam, 1998, p. 27). In this case the phenomenon is explored through historical document analysis. Such documents which trace and record human activity can be regarded as rich data sources to analyse (Mills et al., 2010). In this case two documented accounts of one science-based drama have been analysed, with reference to a wider body of work. In particular the Dr. Lister case is an example of constructivist style pedagogical models associated with the work of Dorothy Heathcote and other drama educator/researchers. Therefore, some foundational concepts and explanations are required before progressing.

Heathcote's Work and Contexts for Active Learning

Dorothy Heathcote is widely regarded as an inspirational and highly skilled teacher who drew upon her early theatre training to develop ways of working dramatically and collaboratively that demonstrated the ideals of many educational theories in practice. There were certainly other practitioners before her who worked with drama strategies in education, going back to the influence of Harriet Finlay-Johnson (1912) who in the early 20th century gained some prominence when she enrolled her pupils as travellers to the North Pole. Finlay-Johnson would take her students out into nature and engage them in real-life learning wherever possible. When this was not possible she sought to replicate these 'real' experiences with the equivalent type of experience in other learning areas as well. It was this vein of thinking that led Finlay-Johnson, and later Heathcote, to a 'dramatic method of teaching' working with students to bring subjects such as history and science 'alive'. What is significant about the Finlay-Johnston, and later Heathcote, work is their realisation that through dramatic processes it was possible to create a fictional but authentic sense of the 'real'. This was to become a fundamental concern for Heathcote, who wanted education not to be regarded as a 'dummy run' for life but concerned with matters of significance and life itself.

Heathcote (2002) identified four main strategies used in her work, beginning with what she saw as the foundation of all her work, 'drama used to explore people'. This is where she often worked as a teacher-in-role. Heathcote realised that working 'inside' the drama along with the children provided huge scope for interacting with them to create opportunities where they could lead and converse about the learning. This approach differs from more transmissive forms of teaching drama and theatre, where teachers may explain or demonstrate a certain approach or technique with students then enacting such, with the teacher an observer or director to the student work. By working inside the dramatic frame with participating students Heathcote was able to not only provide a model of dramatic action but also demonstrate the level of belief and commitment required, and manage the action from within the dramatic frame.

The other three strategies or models that followed were: 'Mantle of the Expert', 'Rolling Role' and the 'Commission Model'. The type of task-based enterprise central to Mantle of the Expert was further explored through later strategies such as Rolling Role and the Commission Model in the 1990s. Rolling Role has similar characteristics to Mantle of the Expert but involves multiple groups working with a common context, with a key feature being the creation and 'publishing' of work, that other groups subsequently work with and 'roll on' to the next. 'Commission model' takes the idea of clients and enterprise to the next level with Heathcote suggesting that authentic 'commissions' be undertaken, with the example of a hospital commissioning the creation of a community garden.

Heathcote's development of Mantle of the Expert emerged in the mid-1970s and this is perhaps her most well-known innovation. The investigative nature of 'mantle' work positions participants within task-based situations where they are endowed with responsibility. The 'mantle' is "one of responsibility or authority by which people carry out their duties" (Heathcote in O'Neill, 2015, p. 115). Furthermore, the participants are deemed responsible for working within an enterprise over a period of time.

These innovative models aim to engage students through grappling with *problems* that stimulate learning. This was achieved through using fictional contexts to bring problems, that would not normally be present to them, into their sphere of experience, with productive tension created through students having to respond to challenges and dilemmas in the active 'now' time of the fictional context.

It is important to note that the notion of 'expert' does not assume that students are to be positioned as specific content experts in a subject area just by calling them such. The choice of role and 'mantle' to begin with should draw on knowledge and skills students already have, or can use. For example, they may be enrolled as medical staff who have to make a choice, based on evidence presented (or find evidence in a living portrait, as occurs with the Carroll/Heathcote version of Dr Lister). Heathcote therefore used non-threatening actions to initiate student entry into the fictional world and build participant confidence and commitment to the drama. The following quotes elaborate on this approach:

> When a child enters a drama, it has to enter another realm of expectancy from the normal, everyday life. It has to have a chance to drag with it not only relevant experience, but that which is relevant to the new elements being dealt with. If we're moving back two centuries, for example, we can't drag the twentieth century back there. We have to drag something else. And yet that something else has to come out of the [current] twentieth century understanding that we have as human beings … The thread you drag over, from which they can begin this process of drama, must be a very common one that doesn't threaten anybody but is, in fact, not ordinary in the way it will be used. (Heathcote, 1988, p. 8)

Heathcote highlighted the importance of teachers using 'threads', actions, concrete materials and everyday actions, with these being repurposed through the process.

The role of the teacher is an empowering role whereby they can play both inside and outside the dramatic frame to lead, guide, coach, and follow in their work with students. In many ways Heathcote and Bolton's work embodies constructivist views of education and Vygotsky's concept of a 'more knowledgeable other' working in a zone of proximal development (Vygotsky, 1978), enhancing what students might be otherwise able to do or know.

> … in the presence of an empowering adult a child can reach beyond his own capacity in carrying out a task. Teacher-in-role enhances this particular adult function. The teacher, through her role, provides a model of high expectations for the enterprise that at first seems out of reach ... In time he has no choice but to aim beyond his normal ability–and to break the confines of rigidly held concepts. (Heathcote & Bolton, 1995, p. 35)

When working within the dramatic frame Heathcote also notes that what is special about working dramatically is that students are able to 'play at life' and live out an experience. They are able to explore ideas and experiences in the 'no penalty' zone of the 'what if' world of drama, they can "test out ideas, try them over again, and generally examine them, without necessarily having to fulfil, in actual life situations, the promises they have tried out in the depicted one" (Heathcote, 1984, pp. 128–129). Notably this piece of writing—'Materials for Significance'—is where Heathcote (1984) discusses the Dr Lister drama and argues the rationale for using drama in the classroom.

She also notes that in choosing what the drama and the work in the classroom should be concerned with, that material should be selected to enable students to investigate 'matters of significance'. In selecting and shaping the learning experience, teachers therefore need to not only consider what needs to be covered from 'the curriculum' but how students will be able to 'interact' with situations through episodes and action. The use of drama is therefore not about 'acting' out scenes from science or history, but using the fictional frame for students to have meaningful experiences in what Heathcote called 'now' time. Much of the type of work Heathcote did has formed the foundations of what is now known as a particular form of drama, Process Drama (Bowell & Heap, 2013; Haseman, 1991; O'Neill, 1995; O'Toole, 1992), whereby teachers and students take on fictional roles and explore ideas and situations within a 'what if' frame. While dramatic strategies may be used, this way of working is often used as a pedagogical process with no intention for theatrical development or presentation.

In the examples of practice this chapter investigates, the dramatic framing works with the people and history of science, recognising that while science may appear to be about 'factual' discipline knowledge that students need to learn, in fact science investigations are conducted by people and impacted by motivations, curiosity and problem-solving. Through working with the 'dramatic' characters of science and dramatic roles, students can learn about science but also the difficulties and dilemmas people faced in the past and how scientific innovations have been achieved.

Key issues that emerge for the teacher/curriculum designer relate to how to engage and facilitate learning through the content and role/s adopted by teachers and students, moving beyond transmissive models of learning to engaged, interactive ones. Another major area of concern relates to how to include and manage working with the science disciplinary knowledge through a dramatic framing. How does dramatic framing work to develop substantial scientific content and skills learning? This is an issue that has been identified previously in work conducted when working with scientific knowledge and recognises the importance of respecting the knowledge base and history of the fields involved (Davis, 2016, 2018; Davis & Tarrant, 2014).

Dr. Lister Accounts

There are several key sources of information about the Dr. Lister drama which have been examined for this chapter. The account that many drama educators would be most familiar with can be found in 'Material for Significance' (Heathcote, 1984) where Heathcote says the drama was developed with a teacher who was working with nine-year-old children. Another key source for analysis is the video that John Carroll created for his doctoral work. His Ph.D. work (Carroll, 1984) was significant for his analysis of framing and the speech language functions and modalities at play throughout the drama. For this work he drew upon the functional linguistics work of Halliday (1973).

The video made of the Dr. Lister work for Carroll's Ph D. submission is also considered a significant innovation for research in the early 1980s, as he included multiple tracks with one focussing on the speech language modalities students were able to use through working dramatically. Of note is the fact he also played the role of Dr. Lister in the drama, with he and Heathcote working as teachers-in-role and adopting what became known as the 'Mantle of the Expert' approach.

Another account which includes considerable detail of this drama is also given in *Joseph Lister: A drama with first school pupils* (Verrier et al., 1986). A published account identifies the authors as a visiting teacher, Ray Verrier who conducted the drama with teacher Audrey Cox and Principal Jean Whitton working with six and seven-year-olds at 'Hampshire First School'. From these sources an outline of the key features of the drama can be determined, whilst also highlighting quite significant differences in the approaches taken.

An analysis of the drama retrospectively has also been written by New Zealand drama educator and academic Peter O'Connor (2013), highlighting some of the key features used in the drama including framing, distancing and the idea of a 'no penalty' zone.

From examining the video and accounts of the Dr. Lister drama in the following section, a description and analysis is made of key decisions regarding the framing and decision-making involved in the drama, with final reflections on some points worth considering for contemporary science and drama-based work.

Heathcote and Framing of the Dr. Lister Drama

Heathcote identifies that the key learning focus for the drama was around students understanding how significant Lister's work was in terms of it being a watershed in medical history (along with and building on the work of Pasteur). While his life in itself is a fascinating one (see profile on Lister later), she highlights the key decision made to 'set that aside' to be able to create a synchronic experience of Lister and the student's contemporary times. The decision is made therefore to find a way to create 'now time' therefore bringing Lister himself into 'our time'.

Children wouldn't therefore be taking on roles that require existing knowledge about the historical era but would look at and 'with' Lister in their own time. The focus should be on the 'outcomes and meaning of material, not action', or involvement say in scientific experiments. Heathcote wanted the children to be able to take on power (which does not mean abdicating teacher power) and influence the construct of meaning. Interestingly she says that 'acting out of stories' places the power in the teacher's hand (Heathcote, 1984). She wanted to look at the inner part of the story, the attitudes of persons, responses from others, and the historical significance of Lister's innovations.

She identifies a range of constructs at play in the framing of the drama. These include:

- Joseph Lister existed in the past, and impacted on that era particularly in the fields of microbiology and surgery;
- Lister's work also impacts on the present time;
- Issues must be seen in the context of Lister's time; but

- The present time must be closely examined; so that
- Comparison of the eras is possible. (Heathcote, 1984)

The way into this complex framing was therefore to enrol the children as students who were studying to be doctors undertaking an examination in the history of medicine. The context, content, and key ideas were introduced initially through setting up a 'living portrait' of Dr. Lister in his office (Fig. 1), with the students enrolled as present-day medical students who had to learn about Lister's work with antiseptics. This arrangement provided a way to bring Lister into contemporary time, and a need for students to learn about the past while still drawing on their present. The children were therefore engaged in learning about aspects of medical science and Lister's work and then documented their learning through writing up their 'findings' as if for an examination.

The video excerpts from John Carroll's tape show a selection from a series of lessons. The initial set up of the drama is depicted through John Carroll, fully costumed as Dr. Lister sitting in the corner of the room. Considerable effort has been exercised to flag that this is not an ordinary classroom setting, with books, pictures, a microscope, surgeon's instruments and certificates suggesting an old-fashioned office of some type. To begin with Carroll, as Lister, is sitting frozen as a living portrait, while Dorothy Heathcote, dressed quite formally in an academic style gown, introduces herself as a medical examining officer. She begins speaking, adopting a sense of authority as she announces that 'I am going to treat you as if you are doctors'. She issues some instructions about what she wants them to observe from the living portrait, specifically 'what kind of Doctor is he' and what they would expect him to be doing. While this may be read as her adopting a high-status teacherly role, John Carroll's narration explains that by Heathcote taking on this role as an authority in opposition, she is actually endowing students with power. This is because she is

Fig. 1 Dorothy Heathcote and John Carroll in the Dr. Lister drama (Screenshot from video)

establishing an expectation that they will succeed, and she is treating them as equals with the responsibilities of 'would-be' doctors. Furthermore, the 'examination' she sets them (to write down what they believe) does not require them to have learnt a body of knowledge, the response invited from them only requires them using direct observation and interpretation of the living portrait before them.

The next frame progresses to a more active role for students as they are invited to ask him questions. This begins to shift the normal classroom speech dynamics through students taking on questioning roles more typical of a teacher. The discipline knowledge work required is this time embodied through the teacher-in-role (John Carroll as Joseph Lister) and the content knowledge expectations for students is minimal but activated through them having to pose the questions which Lister answers.

In later frames of the drama however, students do demonstrate having learnt significant content knowledge. Through using the dramatic framing as the hook, they have over the course of the unit learnt about various 20th century medical developments. Toward the end of the unit students are then able to explain to Lister what modern people now know and the implications of his work. They reassure him that his work was not in vain. Through the various shifts of framing the teacher becomes less of the 'fount of knowledge', and the children gradually *do* become the experts. Carroll and Heathcote used dramatic framing to position the students to meaningfully speculate about possible solutions and futures. Students therefore come to hypothesise and investigate in the ways that scientists do.

While students were involved in science-type experiments, growing moulds and so forth and more typical science type investigations (e.g. with microscopes), the initial entry into the drama required little prior knowledge by the students.

It should be noted that in this version of the drama, the establishment did require a lot of preparation by the 'teachers', in creating the scenario, embodying the content knowledge and expressing such through their fictional roles. What Heathcote argues the value of this preparation and working dramatically can do is "provide the context for bothering to do it" (Heathcote, 1984, p. 134), it provides a means of motivating and engaging students. Later work requires less dramatic framing because she says "that has already done its work" (p. 136).

Other components of the series of lessons outlined in the Heathcote/Carroll work include the following:

- Lister explaining to children his new system for setting bones and sharing his concerns. Students questioning him about his life's work and shared concerns.
- Students finding out what made Lister persist when it was so hard to convince his colleagues of the success of using antiseptic for sterilisation.
- Building understanding of the hazards of work/home conditions in regard to medical processes of that era through children working with historical accounts and images to develop 'waxwork scenes' (or freeze frames) (note: This sequence is featured in the Dr. Lister video and concludes with John Carroll using 'new' technologies by taking instamatic polaroid photos).

- Children work with Lister in the operating theatre as he prepares for his first experiment with the antiseptic spray machine.
- Building an understanding of contemporary developments and the legacy of Lister's work as children teach Lister about modern medical developments through practical demonstrations (kidney machines, blood banks, bank-aids, plasma, aspirin). In the drama they researched different developments and then wore white coats and used medical props such as stethoscopes to introduce him to modern medicine.
- Lister, having explained the significance of receiving his first certificate, (alongside Heathcote as the chief medical officer) presented the children with theirs in a highly symbolic and ritualised presentation ceremony.

Joseph Lister (1827–1912)

Joseph Lister, also known as the 'Father of Modern Surgery' was born in England on April 5, 1827.

His father, Joseph Jackson Lister, was an amateur scientist who designed a microscope lens, with his discovery leading to the modern achromatic (non-colour-distorting) microscope. Joseph Lister was very close to his father, and wrote to him often.

Lister enjoyed experimenting from an early age, doing animal dissections and using a microscope. He was also fluent in German and French, and his facility with languages would later provide him access to cutting-edge research being published in Europe.

Lister initially studied botany and obtained a Bachelor of Arts degree in 1847. He later graduated with a Bachelor of Medicine, entering the Royal College of Surgeons at the age of 26.

In 1854, Lister became first assistant to surgeon James Syme at Edinburgh Royal Infirmary in Scotland, with Symes acknowledged as being a leading surgeon and teacher of the day. Lister later married Syme's daughter, Agnes, who was herself very interested in medicine. Agnes became his laboratory and research assistant and often helped him with experiments and writing up his work. They both loved science so much that on their honeymoon they spent three months visiting leading hospitals and universities in France and Germany. The Listers had no children and so Agnes worked with Joseph throughout his career until her death in 1893. Her contributions were not generally acknowledged in publications at the time, but greater awareness of her work continues to grow.

By 1865 (aged 38), Lister was a professor of surgery at the University of Glasgow. He was concerned that nearly half his amputation cases died from infection. Lister became aware of a paper published by the French chemist, Louis Pasteur and his pioneering work on pasteurization. Speculating upon this work led Lister to believe that the infections might be caused by pollen-like dust carried through the air. Although this was not correct, in 1865 he decided to use a product called 'carbolic acid' to protect the area of operation. Lister tested the results of spraying instruments, the surgical incisions, and dressings with a solution of carbolic acid. He found that the solution swabbed on wounds greatly reduced the incidence of gangrene.

The first patient to benefit from Lister's antiseptic was eleven-year-old James Greenlees. In August 1865, Lister applied a piece of lint dipped in carbolic acid solution onto the wound James had sustained. After four days, the pad was renewed and Lister discovered that no infection had developed. After a total of six weeks the boy's bones had fused back together, without infection. Lister subsequently published his results in *The Lancet*, a leading medical journal.

Lister then instructed surgeons under his responsibility to wear clean gloves and wash their hands before and after operations with 5% carbolic acid solutions. Instruments were washed in the same solution, and the solution was sprayed in the operating theatre. On Lister's wards post-surgery death rates fell to only 15%. While the evidence was substantial it took quite some time for Lister's methods to gain acceptance.

Joseph Lister established his international reputation before his national reputation. He promoted the idea of antiseptic surgery, but nobody in Scotland or England at first believed it. He had to work very hard to provide evidence that his techniques worked. It would take 20 years of experimentation before the medical profession accepted Lister's theory and practical application of antisepsis.

Lister created many other new medical techniques. Among these were naturally dissolving catgut stitches, rubber drainage tubing, a method of repairing kneecaps with metal wire and an improved technique for mastectomies.

Lister died on 10 February 1912 at his country home in Kent at the age of 84.

On the anniversary of his death, in 2012, Lister was celebrated by current doctors and medical scientists as 'The Father of Modern Surgery' (Pitt & Aubin, 2012; Famous Scientists n.d.)

A Different Form for the Joseph Lister Drama: Verrier's Account

The account of the Lister drama by Verrier et al. (1986), shows evidence of some similar ideas to Heathcote/Carroll in the framing but significant differences. Their drama also focusses on the legacy of Lister's contributions, working within historical and current contexts, acknowledging the impact of the past on present medical practices. They also highlight the difficulties involved in change processes. Heathcote's article is referenced, however she is not acknowledged as an originator of the drama. 'Mantle of the expert' and other specific drama conventions are not identified in the learning sequence though the difference between 'telling' the story of Lister and working through story and role play is explicated. Verrier et al. refer to Heathcote to explain the importance of creating a context and 'need' for the investigation. Perhaps surprisingly, they don't introduce the specifics about Lister or tell his story until the very end of the lesson sequence. There is also no teacher in-role to begin the drama or to set up the physical scene as in the Heathcote/Carroll version. This difference in approach is significant in that many non-drama teachers report not feeling comfortable with working with the strategy of teacher-in-role (Balaisis, 2002), so the design of this version of Dr Lister may be useful for examining other ways of working dramatically. Their overview is as follows:

1. Six large pictures are shared with students, showing operations of today and in Lister's time. Children are asked to arrange them into two groups (past and present). They also have to identify what the pictures have in common. They decide if they want to be doctors today or long ago? The majority decided to work in the present. Interest is shown in what modern surgeons wear and so students then draw the dress of modern surgeons.
2. Two pictures from the week are introduced and differences noted. A picture is 'brought to life' with modern student surgeons observing operation by historical surgeon and assistants (the historical surgeon being played by the teacher with five student assistants). At the end of this session, modern student surgeons discuss what would be done differently today. The subject of germs is mentioned but the historical surgeon is confused by what this means. 'Germs' and microorganisms become the topic of discussion and experiments throughout the week using microscopes and magnifying glasses. Sessions involve teacher as Senior Analyst and children in teams of chemists working in pairs. Using scientific investigative approaches, they examined tap water, pond water, sea water, water from a kettle—using their eyes, magnifying glasses and microscopes. This work also led to an evaporation experiment to see what was left once the water had been removed.
3. The surgeon of long ago tells modern students about their worries concerning so many patients dying from septic wounds after their operations. The surgeon then invites modern students to observe an operation and stop at times when actions might be contributing to the wounds turning septic. During this session an incident is introduced featuring one Katy Marston being run over by a horse and cart and rushed to hospital. She is operated on in a very unhygienic fashion and subsequently when they visit her the leg is septic and must be amputated. In-role, the students are positioned as medical students and propose explanations. The students design a hospital to improve conditions and facilities.
4. Real life theological students visiting the school are enrolled as surgeons, nurses and hospital benefactors from the historical era. They talk to the student surgeons and share their opinions. The aim of this experience was for children to elaborate on beliefs of people during historical (Lister's) time and explaining some of the

difficulties of getting people to change. The teacher account indicates children were not as aware of the purpose of this session compared to others. Later on, they wrote up accounts of what they had discussed, and these were added to a wall display.

5. Within the frame of a modern hospital–a version of the ones the students designed–the students assume different work roles. A new head surgeon visits and explains how he wants things changed. His ideas are quite radical, and the children raise objections and discuss the proposals.
6. At the conclusion of the lessons with the visiting teacher (Ray Verrier) he tells them the story of Lister, making links to the experiences they have had. Children write a section of the story and these are put together to tell the whole story.

There are some interesting features of this model and ways that children were able to build their science knowledge and sensibilities and assume ownership of that knowledge. Of note is the use of historical and contemporary images as a way to introduce past and present contexts and begin to build content knowledge. This is instead of the living portrait and medical examiner roles and room set up that Heathcote and Carroll used. The selection of the images is able to activate some similar investigation and observation processes, though the teacher working dramatically role modelling is not here used.

This account also includes some details of actual experiments and use of microscopes, encouraging students to engage in scientific learning through identifying microorganisms in their environments. The historical framing is used to activate more typical science-based investigation and this is also suggested as occurring in the Heathcote/Carroll work. Of note in the Verrier account is the creative use of the visiting adult students and enrolling them in a dramatic framing. This action served to shift responsibility onto the children to articulate their learning through putting their case to the strangers. This therefore endowed them with a 'mantle' of responsibility. It could however be quite a high-risk encounter for some students depending on how secure they felt about what they knew but also in conversing with adults.

It is notable though that at the end of the process, the writing of the story is the way students captured and expressed what they had learnt. It is not clear if dramatic forms were used to present the "story". Some 'active' way for bringing the frame and application of their knowledge back to the present could be added, as these are the types of concerns that Heathcote would have encouraged. Of great significance is the reflection by the principal who had become involved in the drama when the classroom teacher was absent, and her observation that for children of 6 or 7 years of age, their ability to grasp abstract concepts, argue logically and use a scientific approach was surprising.

Summary

The analysis of the Dr. Lister accounts demonstrates that there are multiple ways drama and learning work can be activated to engage and support learning across science and other disciplines. It also highlights the value of learning from historical models of practice, that it is possible and likely that when teachers imagine and create their own versions (as Verrier and colleagues did), it will always be different and that is completely fine.

Similarities between the two accounts include the linking of the past historical context and a present one for the students, and the need to bring the past to life. The goal in both cases is for children to understand the significance of the innovations made by Lister and colleagues, their legacy today, but also the difficulties that confront change agents.

In both cases students were enrolled fairly early on as some type of medical or surgical students and this gives them the impetus and motivation for learning about the relevant medical history, although the early lessons don't ask too much of students in terms of prior science knowledge or dramatic action. They are keyed into the era and challenges of that time and presented with materials that they can investigate. The use of the range of photographs, past and present does appear to be a useful strategy, introducing a lot of information and highlighting comparative aspects early on.

The use of the photographs and accounts of historical surgery practices help reveal the past medical world to the students. The unhygienic practices of the past become evident in comparison to contemporary medical practice and this stimulates students' own investigations into the world of microorganisms.

A key difference in the two accounts can be found in the initial framing, with Heathcote/Carroll's use of teacher-in-role and 'Mantle of the Expert' (Heathcote, 2002), playing Lister and a medical examiner, and the children enrolled as trainee doctors. The second version uses the pictures as stimulus for dramatic work by the children, the teachers initially sit 'outside' the drama as instructors but not participants 'with' the children. The visiting teacher and others do take on roles in later frames of the drama. This approach may be less 'risky' for teachers who are not so comfortable with working as teacher-in-role or embodying the knowledge base required to confidently take on the role of a Dr Lister.

While both accounts acknowledge issues with focusing too much on the Dr. Lister story itself, Heathcote did not hide the fact the drama was focused on his experiences and breakthroughs, whereas the second account hides this fact altogether until the end of the drama. While this could be exciting for children to then discover the parallels, by withholding that information it appears as a lost opportunity for students to further investigate Lister's life and achievements. It does seem clear from the account that students did develop an appreciation for the nature of issues faced by innovators and change agents.

Both accounts end with students 'owning' the knowledge and explaining to others, specifically adults in-role, the significance of various medical breakthroughs. Having children explain to Lister (Heathcote/Carroll version) about the innovations that have occurred since his time, or argue the need for change (the Verrier et al. version), appears to be an effective way of motivating them to learn about micro-organisms and various medical breakthroughs while also developing an appreciation for scientific inquiry.

Conclusions

The Dr. Lister drama was a highly innovative and influential education unit that was used to teach children about the history of medical innovation and specific investigations involving microorganisms and antiseptic. With ongoing calls to make science teaching more engaging, this dramatic framing shows how scientific knowledge and issues can be brought to life through drama, shifting the learning experience from 'learning about' to 'learning within'. The processes used are active and participatory and when the students engage in science-based content knowledge, they are empowered with active and enactive ways to demonstrate that learning. The Dr. Lister drama provides an historical model that was regarded as successful in its day, but given our own times with the reality of pandemics and the ongoing issues related to 'germs', viruses and transmission very much to the fore, a revision of this drama could well be timely and of value. Revisiting and reimaging historical practice still provides instructive advice, and demonstrates effective ways to use dramatic framing to expand contemporary approaches to science education and make learning 'real' and significant.

References

Australian Curriculum, Assessment & Reporting Authority (ACARA). (2016). *The Australian curriculum-science* (Version 8.3). Retrieved from https://www.australiancurriculum.edu.au/f-10-curriculum/science/

Balaisis, J. (2002). The challenge of teaching in role. *Applied Theatre Researcher, 3*(3), 1–7.

Bolton, G. (2003). *The Dorothy Heathcote story: Biography of a remarkable drama teacher*. Trentham Books.

Bowell, P., & Heap, B. (2013). *Planning process drama: Enriching teaching and learning*. Routledge.

Carroll, J. (1984). *The treatment of Dr. Lister: A language functions approach to drama in education* (2nd ed.) [video package]. Bathurst, NSW: Mitchell College of Advanced Education. Retrieved from https://vimeo.com/78321975 (Password: drama)

Davis, S. (2013). The dramatic engineer—The creative life and legacy of John Carroll. *NJ (drama Australia Journal), 37*(1), 41–52. https://doi.org/10.1080/14452294.2013.11649562

Davis, S. (2016). Learning that matters: Revitalising Heathcote's rolling role for the digital age Rotterdam, The Netherlands: Sense Publishing.

Davis, S. (2018). The engagement tree: Arts-based pedagogies for environmental learning. *IJEA (International Journal of Education and the Arts), 19*(8). Retrieved from http://www.ijea.org/v19n8/v19n8.pdf

Davis, S., & Tarrant, M. (2014). Environmentalism, stories and science: Exploring applied theatre processes for sustainability education. *RiDE (research in Drama Education), 19*(2), 190–194. https://doi.org/10.1080/13569783.2014.895613
Famous Scientists (n.d.) *Joseph Lister.* Retrieved from https://www.famousscientists.org/joseph-lister/
Finlay-Johnson, H. (1912). *The dramatic method of teaching.* Boston & New York: Ginn & Company.
Halliday, M. A. K. (1973). *Explorations in the functions of language.* London, UK: Edward Arnold.
Haseman, B. (1991). Improvisation, process drama and dramatic art. *London Drama*, 19–21
Heathcote, D. (1984). Material for significance. In L. Johnson & C. O'Neill (Eds.), *Dorothy Heathcote: Collected writings on education and drama* (pp. 126–137). Northwestern University Press.
Heathcote, D. (1988). *Aspects of drama for learning: Dorothy Heathcote talks about sequencing, planning, styles of teaching and other topics.* Newmarket, NZ: Kohia Teachers Centre.
Heathcote, D. (2002). *Contexts for active learning—Four models to forge links between schooling and society.* Paper presented at the NATC, Birmingham. Retrieved from http://www.moeplanning.co.uk/wp-content/uploads/2008/05/dh-contexts-for-active-learning.pdf
Heathcote, D., & Bolton, G. (1995). *Drama for learning: Dorothy Heathcote's Mantle of the Expert approach to education.* Heinemann.
Merriam, S. B. (1998). *Qualitative research and case study applications in education.* Jossey-Bass.
Mills, A. J., Durepos, G., & Wiebe, E. (2010). *Encyclopedia of case study research* (Vols. 1–0). Thousand Oaks, CA: SAGE Publications, Inc. https://doi.org/10.4135/9781412957397
O'Connor, P. (2013). The Joseph Lister drama: John Carroll from a distance. *NJ, 37*(1), 7–13. https://doi.org/10.1080/14452294.2013.11649559
O'Neill, C. (1995). *Drama worlds: A framework for process drama.* Heinemann.
O'Neill, C. (2015). *Dorothy Heathcote on education and drama: Essential writings.* Routledge.
O'Toole, J. (1992). *The process of drama: Negotiating art and meaning.* London & New York: Routledge.
Pitt, D., & Aubin, J. M. (2012). Joseph Lister: Father of modern surgery. *Canadian Journal of Surgery/journal Canadien De Chirurgie, 55*(5), E8–E9. https://doi.org/10.1503/cjs.007112
Verrier, R., Cox, A., & Whitton, J. (1986). Joseph Lister: A project with first school pupils. *Education 3–13, 14*(1), 45–50. https://doi.org/10.1080/03004278685200111
Vygotsky, L. S. (1978). *Mind in society: The development of higher psychological processes.* Harvard University Press.
Yin, R. K. (2009). How to do better case studies. In *The SAGE handbook of applied social research methods* (pp. 254–282). SAGE Publications. doi:org.ezproxy.cqu.edu.au/https://doi.org/10.4135/9781483348

Susan Davis (B.A., G.D.Teach., G.D.Admin., M.A. Research, Ph.D.) is an Adjunct Associate Professor with CQUniversity with extensive experience as an arts educator, researcher and practitioner. Her research has focused on drama, engagement, digital technologies, cultural-historical theory, creativity in education and sustainability. She has had five books and over 50 book chapters and refereed articles published, including the book "Dramatic interactions in education: Vygotskian and sociocultural approaches to drama, education and research", and "Learning that matters: Revitalising Heathcote's Rolling Role for the digital age".

Dramatising the S and M in STEM

Kathryn Paige, Leni Brown, Lisa O'Keeffe, and Robyne Garrett

Abstract Interdisciplinary science and mathematics learning can be enhanced via the dynamics of drama in the form of creative and body-based learning (CBL). This pedagogical approach allows pre-service teachers to be engaged personally, to express their own ideas and to help students discover meaningful connections to mathematical and science concepts. A professional learning program that was generated through collaboration between a group of South Australian and Texan academics in the areas of mathematics and arts highlighted the potential of CBL to provide new and innovative pedagogical approaches to the teaching of science and mathematics curriculum. This chapter documents strategies in which science and mathematical concepts form the basis of interactive scenarios, role-plays and drama games that offer pedagogical alternatives for teachers and pre-service teachers to engage their own students in developing conceptual understanding. Practical strategies which have been modelled with pre-service teachers, primary teachers and their classes are described. These strategies include: 'The Truth about Me'; 'Alphabet Relay'; 'People, Shelter, Storm'; and the sum of internal angles through the 'Human Geoboard' and the 'Sir Cumference Quest'. Dramatising gears, electrical circuits, dichotomous keys and equitable distribution of world resources allows a focus on particular science and mathematics concepts.

Keywords Creative body-based learning · Science · Mathematics · Professional learning · Drama · Pre-service teachers

K. Paige (✉) · L. Brown · L. O'Keeffe · R. Garrett
University of South Australia, Adelaide, South Australia, Australia
e-mail: kathryn.paige@unisa.edu.au

L. Brown
e-mail: leni.brown@unisa.edu.au

L. O'Keeffe
e-mail: lisa.okeeffe@unisa.edu.au

R. Garrett
e-mail: robyne.garrett@unisa.edu.au

P. J White et al. (eds.), *Science and Drama: Contemporary and Creative Approaches to Teaching and Learning*, https://doi.org/10.1007/978-3-030-84401-1_5

Introduction

One of the most significant parameters determining the direction and quality of classroom learning (Ball, 2000) is teachers' pedagogical practice, so this must be reflective and inclusive of the diversity of learners in classrooms today (van Kraayenoord, 2007). If there is no focus on the formulation of more effective teaching policies and practices to improve classroom learning, there will be no significant decrease in the number of students who become alienated from the educational process each year. Hence, schools must continually develop practices that are more inclusive and responsive to student diversity. For science and mathematics education the challenges of inclusive and responsive teaching are both complex and significant. For example, Sleeter (2012) highlights how the political nature of education slows development in mathematics education in particular; where a focus on basic skill instruction triumphs over sociocultural approaches that are both culturally and contextually rich. Science and mathematics teaching require the development of multiple opportunities for concept building, relevant challenging questions, problem-solving, reasoning, and connections between the curriculum and real-world situations. Mathematics teaching often relies too heavily on rote memorisation isolated from meaning, which represents a narrow view of mathematical foundations (Sherman et al., 2015), while science is still often seen as an optional extra, not embedded in regular classroom programs. For example, in 2013 the Chief Scientist (Chubb, 2013) indicated that the amount of time being allocated to primary science was declining, with less that 3% of weekly instruction time being devoted to primary science in comparison to 18% for mathematics.

There have been some progressive programs in primary schools that have supported innovative teaching in mathematics and science. The Connected Mathematics series is one example of well researched inquiry-based learning that focuses on problem-solving. There are also many professional learning approaches that develop teachers' ability to teach mathematics deeply including: Australian Association for Mathematics Teachers (n.d.) 'Maths300' which has a focus on thinking mathematically, Baker and Baker's (2019) 'Natural Maths', and Siemon et al. (2012) 'Big Ideas in Number'. For science, the content strands from the Australian Curriculum, Science Understanding, Science Inquiry Skills and Science as a Human Endeavour (Australian Curriculum, Assessment and Reporting Authority [ACARA], 2017) as well as the Primary Connections program (Australian Academy of Science [AAS], 2020) continue to be well resourced and influential in improving the teaching of primary science in Australia.

Programs and resources such as those listed above attempt to make mathematics and science more connected to the real world of the students and support teachers, to move away from traditional teacher-centred, pen and paper approaches, as well as making science a central part of the primary curriculum. The common intention of such initiatives has always been to make the context of the mathematics and science content relevant and meaningful to students. This has been done through the application of content to authentic problems that connect to the real worlds of the students

(Brown et al., 2017; O'Keeffe & Paige, 2019; Paige et al., 2019; Zevenbergen, 2000). The premise is that relevance creates accessibility for the students.

In addition to using authentic experiences and problems, another way to support students to build meaningful connections to content is to use embodied pedagogies (Garrett & Wrench, 2016; Koch, 2006; Yoo & Loch, 2016). Drawing on ideas around the body's capacity to affect and be affected (Massumi, 2002); the body as a pedagogic device (Ivinson, 2012); and embodied pedagogy (Nguyen & Larson, 2015), this work focuses on the body's power to feel, sense, respond and imagine, thus its capacity to construct meaning. Engagement with these ideas represents a pedagogical shift towards understanding the body as a vehicle to effect change and allow access to learning (Garrett et al., 2018). While early educational theorists have long advocated engagement with the body in schooling (Dewey, 1938), modern schooling practices often discourage movement in the classroom and deny these kinds of learning outcomes (Berdayes, 2004; Bresler, 2004; Stevens, 2012).

Educators have also argued for the integration of arts-based pedagogies into the core curriculum in order to provide deeper, richer learning environments (Robinson, 2013). Arts-based pedagogies draw on one or more art forms to assist students to deepen their understanding in non-arts learning. They also engage students in acts of metacognition where reflection is drawn to the 'thinking behind' their responses, from multiple perspectives. Importantly, arts pedagogies provide multiple visual, active and dialogic modes to express knowledge, providing both opportunities and access for those with barriers to historically accepted academic literacies (Hirsh, 2010).

Creative and Body-Based Learning

Drama-based pedagogy is a pedagogical approach developed by Katie Dawson from the University of Texas that draws on critical approaches from Boal (1974) and Freire (1996). The practice uses active and dramatic approaches to engage students in embodied, affective and creative practices in order to increase engagement in and access to academic learning (Dawson & Kiger Lee, 2018). Specific strategies include 'activating dialogue' strategies, which utilise verbal, written and embodied dialogue to connect prior knowledge to new ideas. 'Game as metaphor' utilises bodily awareness as well as social and group skills to rehearse understandings. 'Image work' strategies use the body or objects to develop concrete representations of abstract ideas and 'role work' activities invite students to act out roles in problem-solving scenarios (Dawson & Kiger Lee, 2018). Also integral is a process of dialogic meaning making that aims to develop understanding through interactive exchange and opportunities to *describe* actions, *analyse* learning and *relate* to student life-worlds.

Our interpretation of the conceptual resources offered by embodied and drama-based pedagogy has led to the development of a creative and body-based learning (CBL) initiative in Adelaide, South Australia. In expanding the specific strategies of drama-based pedagogy to include multiple forms of embodied and artistic practice, the CBL curriculum approach is intended as a pedagogical *provocation* for teachers to reimagine pedagogical practice in ways that utilise the body and creativity. The aim of CBL is to engage students on multiple levels: physically, cognitively, socially and emotionally. Its approach to mathematics and science involves learning that is experiential and utilises problem-solving and critical and creative thinking, all of which link well to the general capabilities in the Australian Curriculum. Studies have shown such approaches to be effective in promoting student engagement (Garrett et al., 2018), achievement in geometry (Duatepe-Paksu & Ubuz, 2009), achievement in other specific areas of mathematics (Fleming et al., 2004; Walker et al., 2011), as well as culturally responsive pedagogy (Hardiman et al., 2014).

For this chapter we draw on CBL strategies developed by Katie Dawson and the four authors within a two-year professional development project that aimed to offer new alternative pedagogical approaches to teachers and pre-service teachers in rich scientific and mathematical thinking with which they can engage students. Our aim is not only to consolidate conceptual understanding but to improve engagement, disposition and access to important mathematical and scientific concepts.

Creative Body Learning in Teacher Education

Three authors with a mathematics and science background and one with a dance background were integral in running the professional learning workshops with teachers over a two-year period. We also adapted and implemented various strategies with pre-service teachers in the science and mathematics primary curriculum courses and a final-year pathway into the profession course.

CBL strategies, such as those described below, can be used to accentuate the A in STEAM (Science, Technology, Engineering, *Arts* and Mathematics) while offering an alternative means to engage students in rich mathematical and scientific thinking (the S and M in STEM). Our aim in promoting CBL approaches as a pedagogical vehicle for science and mathematics is not only to support teachers to consolidate students' conceptual understanding, but also to improve engagement and enjoyment in science and mathematics and to encourage students to develop a more positive disposition towards mathematics and/or science. As a research team, we include academics with expertise and a background in science and mathematics, eco-justice, arts integration and embodied pedagogies. Our shared commitment to social justice in schooling frames our focus on 'engagement' as a central feature of transformative pedagogy (Callow & Orlando, 2015). Over a span of 3 funded projects, 40 teachers engaged in action research projects that explored the impact of CBL strategies on their students' learning, such as its impact on disposition around mathematics, engagement of Indigenous students and in response to education needs of students impacted by trauma (Garrett, 2018).

We believe that a combination of CBL and direct/traditional approaches to the teaching of science and mathematics teaching has great potential to promote dialogue, engagement and access to science and mathematics for children. In the work that follows we provide outlines of eight CBL strategies combined with mathematics and science curriculum content. The first four examples focus on CBL and mathematics, the fifth example integrates mathematics and science, while examples six and seven focus on science. The final example is based on a transdisciplinary issue, exploring a fairer world.

Example 1: Alphabet Relay

Prior to beginning a unit of work, it is important to elicit the learners' prior knowledge and to consider what aspects the students will need to understand to fully participate in the following lessons. Consideration should be given to the diversity of learners and how best to help each of them either access, or impart, this information (Van de Walle et al., 2015). The following prior knowledge learning experience could be employed as an authentic assessment tool (Meloney, 2016). The Alphabet Relay (see Fig. 1) is an example of an initial engagement activity (Dawson & Kiger Lee, 2018) or a prior knowledge learning experience (Meloney, 2016), where students work in relay groups to try to provide words that relate in some way to the key concept. The example in Fig. 1 is based on the word 'angle'. Starting with a list of each letter of the alphabet, students take turns to assign their words sequentially, that is, they must fill in responses in order from A to Z. The first group to complete the alphabet are the 'winners'. Importantly, students work in teams and can help and give ideas to each other. A key component of this activity are the rich discussions that can be held afterwards, which can include discussions about the frequency of some terms as well as understanding of different terminology, the identification of words that some may not agree with and the common 'blanks'. Such discussions provide an opportunity to activate student thinking about the focus topic.

Alphabet Race

Group 1 (8 children)	Group 2	Group 3
Acute	*Angle*	*Arc*
Bisect	*Bearings*	*B*
Convex	*Corner*	*Complementary*
Degrees	*Diameter*	*D*
Equal	*Equivalent*	*Exterior*
F	*F*	*F*
Geometry	*Grade*	*G*
Half turn	*H*	*H*
I	*Intersect*	*I*

Fig. 1 Alphabet relay: this example is based on language related to the word 'angle'

Example 2: The Truth About Me

'The Truth about Me' (Dawson & Kiger Lee, 2018, p. 140) is another example of an engaging game-based learning strategy. It begins with the teacher inviting students to stand in a circle with a small marker on the ground (masking tape) for each student and the teacher in the middle. The activity is introduced by saying we are going to try to find out a little more about everyone. The person in the centre will share something about themselves by saying, '*The truth about me is …'* They complete the sentence with a statement that is true for themselves, for example: *'The truth about me is that I have a sister'*. All students in the circle for whom this statement is also correct must move from their spot and find a new spot around the circle. The person in the centre also moves to a spot in the circle, leaving the centre vacant for the last person without a spot. The game then begins again with a new person and a new truth statement. A good facilitation strategy is to encourage the use of statements that are likely to be true for a lot of people and it is important to play a number of times to develop familiarisation and fluency. Once this is established, the rules for the 'truth' can shift to relate to an aspect of mathematics or science. An example is giving everyone a label or a picture of a different vertebrate or invertebrate and when a grouping is called out such as 'insect' those who are insects move. Similarly, with mathematics one could use shapes (quadrilaterals, etc.) or numbers less than 30, using properties such as odd, even, square or perfect.

Example 3: People, Shelter, Storm

'People, Shelter, Storm' (Dawson & Kiger Lee, 2018) is an example of a theatre game learning strategy. It can be introduced as a group activity to familiarise students with the technique. Initially, the whole group simply moves around a space. On a cue from the teacher students form groups of three with two of the trio creating a 'roof' with their arms and the third person sheltering under it. From this point onwards when the word 'shelter' is called, students who have formed the shelter release their arms and find a new partner to form another shelter above a 'sheltering' person. When 'people' is called out, the shelters stay on their spots and the people have to run to another shelter. The final word is 'storm', where everyone leaves their position, people can become shelters, shelters can become people, or they can decide to stay the same. They can also move their body like a storm. Once students are familiar with the strategy, mathematical content can be added. Groups can be instructed to form a variety of angles when the word 'people' is called out (see Fig. 2), with an emphasis on an accurate representation. While the angles may not be mathematically accurate, they do provide a clear insight into student understanding of the various angle attributes.

Fig. 2 People, shelter, storm: this example is focused on forming and describing angles

Example 4: Sir Cumference of Angleland

The concept of an angle can be further developed by groups creating different angles using a piece of string. For example, the string can be used to create two rays, with one person as the pivot point and the other two holding on to the ends of the rays. This is also a way to develop student understanding about dynamic angles, as the rays have to move to create the angle. A good discussion point for this task is questioning how the group created their angle, which can focus students' thinking on the idea of the angle being the 'amount of turn' of the rays. This technique also has the capacity and flexibility to allow greater depth of understanding of student prior knowledge. Further extension can be achieved by making the string into a loop (e.g. knotting the ends) then asking the groups of three to use the string to create different types of triangles. The properties of each triangle would determine the angle size and length of sides, and so the emphasis would be on the group understanding of the differing properties of each triangle.

Through reading an excerpt from the mathematical tale 'Sir Cumference and the Great Knights of Angleland' (Neuschwander, 2001) as a starting point, a context is given to the ensuing activity which can be used to further develop students' understanding of the geometric properties of angles. A quest is bestowed upon the chief knight, Sir Radius, to find a solution to a dilemma being faced by the Queen of Angleland, who wants to create a new jousting arena for her 12 Royal Knights. The Queen's challenge to Sir Radius is to explore how all twelve Royal Knights of Angleland can be allocated an equal amount of area in the new dodecagon shaped jousting complex.

The scene is set to use role-play. The leader of the session should step into the role by dressing in a cape or hat and assuming the mantle of the Queen's chief knight, Sir Radius. The class is invited to join the quest and become 'Knights of Geometer'. Protractors threaded on string are bestowed around each student's neck as their badge of knighthood and their task is clearly stated: to help the Chief Knight

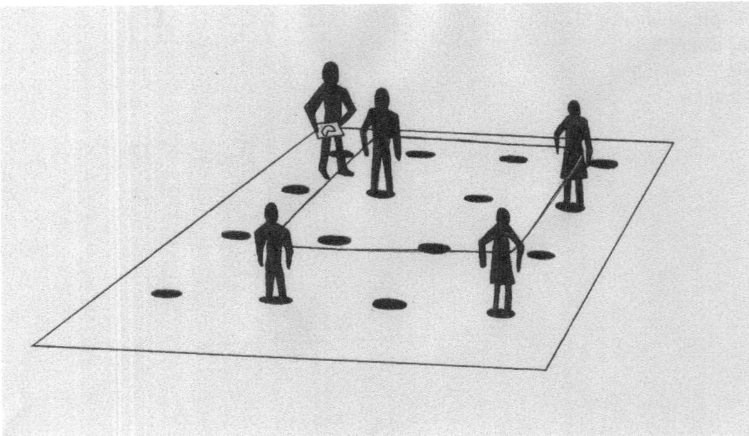

Fig. 3 An example of a giant geoboard

solve the Queen's dilemma. Sir Radius leads the quest through the exploration of the relationship between the sum of the interior angles of a polygon and the number of its sides.

It can be established, through prior knowledge, that triangles seem to have a constant 180 degrees as the sum of their interior angles, regardless of their shape or size. This can prompt explorations such as: *'What can be discovered about the sum of the interior angles of quadrilaterals?'* A giant geoboard can be created using masking tape on the floor to construct quadrilaterals (see Fig. 3). On this geoboard the students can be the pegs and balls of string can be used instead of rubber bands to create the shapes.

A variety of quadrilaterals can be constructed and recorded (on paper or whiteboards) and different groups of students can be challenged to determine and calculate the total interior angle sum using rotagrams and protractors. What should be established is that, regardless of the quadrilateral shape, that is, whether it is a rhombus, trapezium or square, the angle sum is always 360°. Once this is established, the following question, to challenge mathematical thinking, could be posed by Sir Radius: *'If triangles have a constant sum of their interior angles, and now that we now have established that quadrilaterals also have a constant sum of their interior angles, does the same apply to other polygons?'*

Each small group can then be allocated a specific polygon, either a pentagon, hexagon, octagon, nonagon or a decagon, and tasked to construct 'regular' and 'irregular' examples on their geoboards, recording the shapes created (on paper or whiteboards) and determining the interior angle sum. The results for each group can be recorded on a collective table to establish an emerging pattern and the relationship between the number of sides of a polygon and its interior angle sum. The students can then be challenged to identify these patterns and relationships for themselves and possibly the number of triangles that can be created from each shape.

The knights will be successful in their quest if they are able to use this gathered information to identify an emerging pattern and the relationship between the number of sides and the sum of the interior angles. They can continue the pattern to determine the sum of the interior angles of a dodecagon, and therefore the size of each interior angle of a regular dodecagon, and hence provide an answer to the Queen of Angleland's original quest. They can also be challenged to generate an algebraic expression that could be presented to the Queen to ensure each knight has an equal area in the new jousting complex.

Example 5: Curator Role-Play, Sorting and Classifying in the Museum and Dichotomous Key

This learning strategy revolves around an 'everyday' exhibit being set up by a national museum and has explicit links to both mathematics and science curriculum. 'Katie Curator' has been given the task of organising an exhibition for the museum involving displays of everyday items. A collection of objects, such as Lego, rocks, feathers, Bratz Dolls' clothes and buttons, are put on display. How these are to be set up is the responsibility of the teacher or pre-service teacher. In groups of 3, students are tasked with creating an interesting display. To add some sense of order they can first be asked to generate a list of 10 different attributes or properties they could use to sort and classify their collection. Students then choose one of these attributes and use it to create their display with a creative and original title. This title should reflect the 'mode of classification' or the objects, or both. Following a gallery walk to review each other's displays and discussing how each group sorted their objects, the head curator (group leader) can check on the progress and then inform the groups that the museum wants to take their artefacts on a road tour. This presents the problem of how to faithfully recreate their displays at each new venue, thus creating a need to develop a dichotomous key. Students are then tasked with creating their own dichotomous key, using their museum display pieces as the basis. Before sending students off to create this in their groups it is advisable to work on a class dichotomous key as a group. An example using shoes and masking tape to create a human dichotomous key is shown in Fig. 4. Additionally, a giant dichotomous key could be drawn outside, for example to classify rubbish found in the schoolyard, using such properties as recyclable and non-recyclable.

The following are examples of CBL and science topics. In developing these examples, we have been influenced by the ideals of the McClintock Collective (1988), a group of educators who focused on developing gender-inclusive teaching strategies in science.

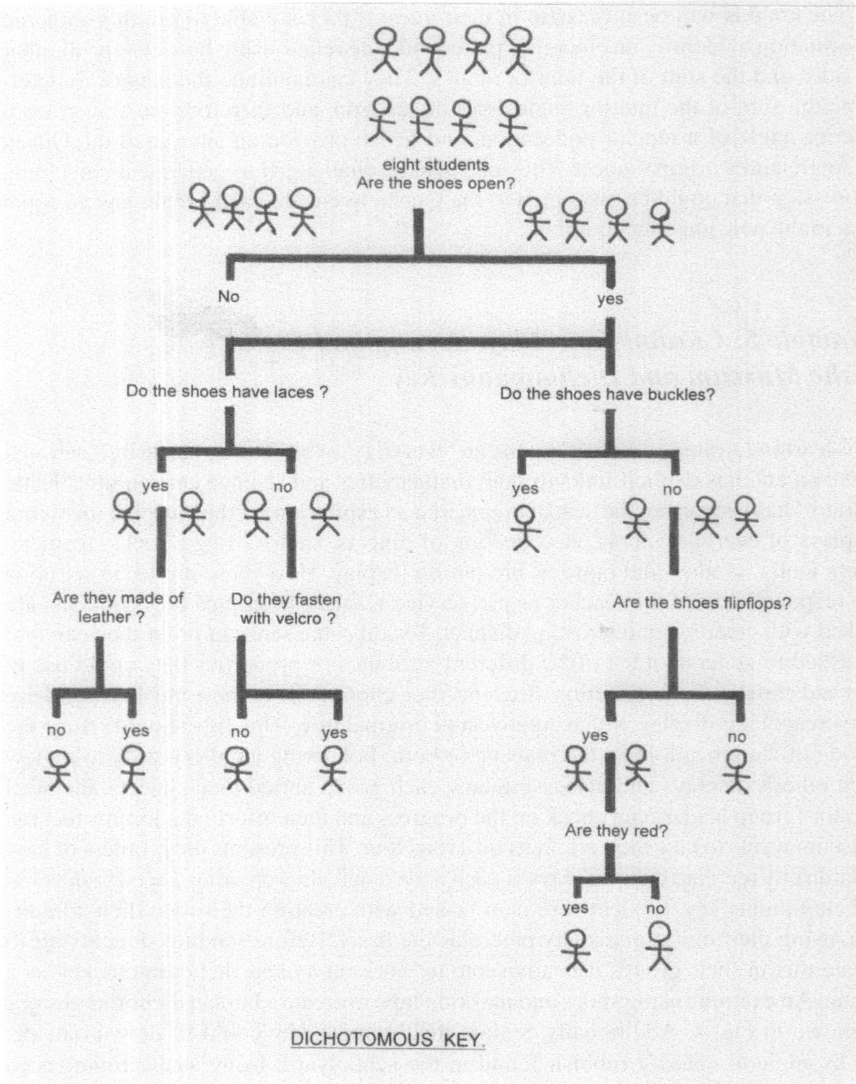

Fig. 4 An example of a dichotomous key

Example 6: Cogs, Gears and Simple Machines

'Cogs, Gears and Simple Machines' provide rich opportunities for STEM exploration. However, the comparison of the ratio of the number of cogs in a cogwheel to the number of revolutions or even 'what is a gear' can prove to be very challenging

Fig. 5 Example of cogwheels connecting

concepts for children to understand. Using drama to investigate cogs and further develop the idea of gear ratios provides an interactive vehicle where everyone is involved. This can help to consolidate the learning in a context where students can embody and embrace some of the key ideas. The simplicity of the drama approach is deceptive in the power that it holds to bring together the relationship between science and mathematics in a way that enables children to understand its importance (McClintock Collective, 1988).

The activity begins with a group of students being divided into two groups, one large group and one small group. For example, if there are 25 students you could have a group of 20 and a group of 5 (see Fig. 5). The larger group forms the first of two circles, facing forward, where each student puts their left arm on the shoulder of the person in front and extends their right arm out to the side. This represents the cogwheel and its teeth (cogs).

To create the outline of the cogwheel and to ensure students have a clear path to follow when turning, a circle can be marked out on the ground using chalk, string, rope or anything else available. The smaller group of students is also arranged in a circular configuration, but with their right arm connecting to the person in front. Both circles, or cogwheels need to connect and so a link, point 'A', is established, as seen in Fig. 5. The students at this point need to be identified at the beginning of the revolution. The large cogwheel now can begin moving slowly around its circle, ensuring that each 'cog' (or raised arm) fits into the space created by two 'cogs' on the small cogwheel. Once the link person from the larger cogwheel is back in their original spot, having completed one full revolution, check with the link person in the smaller circle as to how many revolutions the smaller cogwheel has completed. The size of the cogwheels can be varied; however, it is a good idea to aim for integer ratios such as 24 + 6 or 21 + 7. Students should be asked to predict what will happen if they change the speed or direction of the cogwheels. They also need to experience being part of the large cogwheel as well as the smaller cogwheel so they can begin to make connections to the difference the size of the cogwheel and the number of cogs makes to the *speed of the revolutions*. Some good questions for discussion include: '*What is moving quicker: the larger cogwheel or the smaller?*', '*How many more turns does the larger cogwheel do compared to the smaller wheel?*', '*What is 'gear ratio' or how does the turning of one wheel impact on the other?*', '*Where do we see cogwheels in everyday life?*' All of these questions are intended to reinforce the importance of the role of cogs in a machine and all have the potential to be answered through this type of dramatic enactment.

Example 7: Electrical Circuits

Electricity is often considered to be a mysterious concept. Physically acting out scientific concepts that are challenging to understand in a concrete manner is a useful approach. It allows a deeper and embodied understanding of abstract processes, which are especially prevalent in chemical and physical concepts such as electricity flow (McClintock Collective, 1988). This approach provides an alternative learning strategy that may support students who have difficulty understanding concepts from more traditional approaches such as textbook explanations or experimentation. It allows students to place themselves in a position so that they can see or feel components as well as experience how they interact together.

To develop the idea of the flow of current through electrical wires, the following props are needed: a box labelled battery, a collection of hats labelled +ve terminal, −ve terminal and switch, enough +ve charge cards, denoting the units of positive charge, for each student, the same amount of −ve charge cards presenting units of negative charge, or electrons, with safety pins attached, some extra −ve cards to place in the battery box (dry cell), and some rope or string. Students could be involved from the very beginning in making this equipment.

To begin, one student is appointed as the battery and the rope/wire is placed in a circle, representing the wire in a circuit. All other students are to pin a +ve charge card onto their fronts, to represent the 'positive holes' in the circuit and then stand equally spaced along the rope, holding it in their left hand, leaving their right arm free to pass the 'charge' (see Fig. 6). Two students are appointed as the positive and negative terminals, denoted by their hats, and they are to position themselves either side of the battery, with the negative terminal to the left of battery and the positive to the right. The battery box (or dry cell) containing the extra −ve cards is place in front of the terminals and the 'battery' student is the keeper of the charge as their role is to continually supply the negative terminal with −ve charges. Each student, except the one standing next to the negative terminal, is then given a −ve card and asked to hold this over their +ve card, to signify electrical neutrality.

Before the 'enacting' begins it is important to elicit student thinking about how they think the role-play should proceed to show the flow of the negative charge. For example, have them consider and discuss the flow of the −ve charge from the negative terminal to the positive terminal and how this would happen, or where the

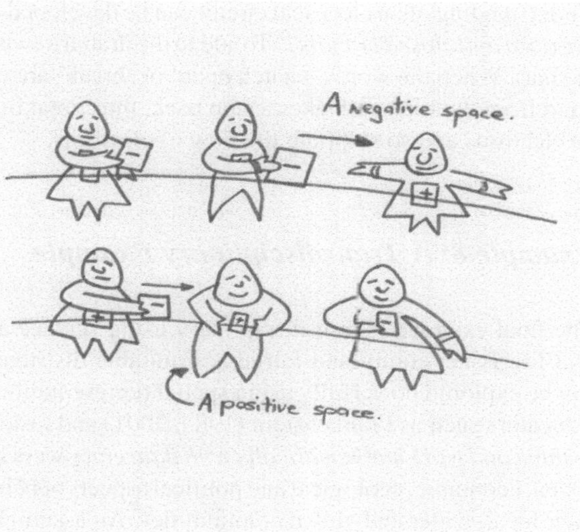

Fig. 6 Example of circuit role-play

charge will come from or go to, or what will be the speed of the flow and how this might be determined. Some may already have read a textbook definition, and this may challenge them to translate the theory into practice.

The following rules need to be established before the enacting begins. A student can only receive a −ve card if they are a +ve gap or 'hole'. This is to represent the flow of current along a wire. The cards can only be passed and received with the free right hand in a clockwise direction. Once the −ve card is accepted it must be placed in front of the +ve card to symbolise the neutral part of the wire. The 'battery person', as the keeper of the charge, is the one who will regulate the rhythm and can determine what this will be by indicating when the charge will flow. They are also responsible for supplying the negative terminal student with a −ve when a +ve hole is created at the conclusion of a completed circuit of charge flow. It is important to emphasise the most energy-efficient way for the battery student to supply the charge. They must expound energy to take out the −ve, or the charge, and lift it from the box, whereas the positive terminal can simply drop the card into the battery box, an action requiring little energy. This emulates the reactions occurring in a battery that 'push and pull' the electrons. The role-play should be continued for at least three revolutions, with the −ve charge flowing and allowing the students to see the +ve hole flow and to develop a feel for charge flow.

To reinforce the concepts, questions need to be posed and answered, such as: *'What was the direction of the negative charge (−ve terminal to +ve)?'*, *'What is the flow of a conventional current (+ve terminal to −ve)*?', *'What comparisons can be made between how the battery person was lifting and passing the electrons and what is going on at the terminals of a battery?'* The participants are replicating what chemicals in a battery do when they apply the 'push' to the electrons, as they must be passed with energy. *'How could you draw a circuit to represent the drama?'*

Understanding of an electrical circuit can be developed by posing the question: '*What are open and closed circuits?*' To add to the drama a switch could be introduced (using the hat). When the words 'switch open' or 'break' are used, the switch removes their hand from the rope and takes a step back, thus breaking the ability to hand on the –ve electrons and so stopping the flow of electrons.

Example 8: A Transdisciplinary Example

The final example is transdisciplinary using science and mathematics to explore a real-life issue, equity and fairness. Equitable distribution of the world's resources can be explored powerfully using spatial representation and data handling strategies. Educators such as Frankenstein (1983, 2001) and Osler (2007) argue that the ability to *think and work mathematically with data* empowers students to participate in many social, economic, ecological and political aspects of being a responsible citizen within their local, wider and global communities. An assumption here is that one of education's key roles is to develop responsible, critical citizens, which in turn suggests the need for a mathematics education for democracy (for example, data handling has to be more than drawing a graph of meaningless data and answering five questions about the graph). From a mathematical perspective, Chartres (2008) suggests that mathematics can be used in ways that empower students, connecting with big ideas such as citizenship and democratic participation, but it can also be used to disempower and reinforce marginalisation and/or exclusion. Critical mathematics education and ethnomathematics provide opportunities for more equitable experiences of mathematics (O'Keeffe & Paige, 2019). This transdisciplinary learning experience uses bodies to represent the inequity of resource distribution and provides opportunities for students to think about the impact their ecological footprint has on the world.

The learning experience begins with the introduction of the game called 'Shelter, People, Storm', where the teacher divides the students into random groups of three (see Example 2). Parts of the video 'The story of stuff' (Leonard, 2009) are then viewed to engage participants in thinking about the power of data to convey messages about the environment, the world's population and consumption. However, for young people it is very difficult to relate to such global issues because of the enormity of the data that is represented. 'If the world was a village of 100 people' (Master Communications, 2010) is a useful embodied strategy to represent data in a more manageable way. Using the $\frac{4}{5}$: $\frac{1}{5}$ model (outlined below) is another way.

Using masking-tape, a 4-m square, divided into 5 equal rectangles as shown in Fig. 7, is created on the ground. This is to represent a very general $\frac{4}{5}$: $\frac{1}{5}$ model, often used to illustrate how the planet's resources and wealth are 'owned' and consumed. Put simply, $\frac{1}{5}$ of the world's population access, 'own' and consume $\frac{4}{5}$ of the planet's resources. It is a powerful way to engage a person's critical lens. The square represents the planet's total wealth and a class represents the world's population. $\frac{4}{5}$ of the class then attempt to occupy one of the rectangles ($\frac{1}{5}$ of the square without going outside of

Fig. 7 Example of a $\frac{4}{5}$: $\frac{1}{5}$ model (a 4-m square, divided into 5 equal rectangles)

the boundaries), whilst the other $\frac{1}{5}$ of the class occupies the other $\frac{4}{5}$ (or 4 rectangles) of the square. In a class of 25, twenty students would occupy the shaded $\frac{1}{5}$ of the square and 5 would occupy the remaining $\frac{4}{5}$ of the square (i.e. the other 4 rectangles).

Whilst the mathematics is quite simple, the message that it conveys is very powerful, and easily understood by children. It can be taken further and repeated over and over again, with each rectangle being divided into fifths, so that each square represents $\frac{1}{25}$ (see Fig. 8).

If the model is followed to the second step, $\frac{4}{5}$ of the people (students) who occupy the original $\frac{1}{5}$ of the square (the shaded rectangle) access and consume $\frac{1}{5}$ of the world's resources and wealth. In a class of 25, there are 20 students already occupying this rectangle and this would leave 16 students occupying $\frac{1}{25}$ of the large square. Similarly,

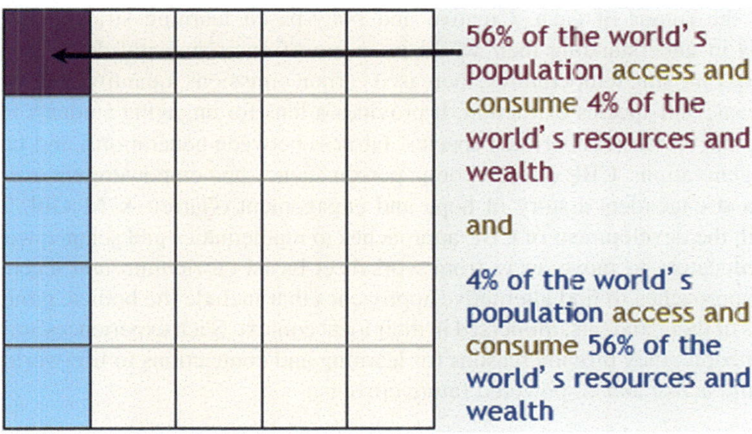

Fig. 8 Second iteration of the $\frac{4}{5}$: $\frac{1}{5}$ model

at the other end of the scale $\frac{1}{5}$ of $\frac{1}{5}$ of the class, ($\frac{1}{5}$ of 5) can access and consume $\frac{4}{5}$ of $\frac{4}{5}$ of the original square. That is, one student can occupy sixteen of the 25 smaller squares.

Some supporting questions to guide this activity include: '*Which issues often highlighted or ignored in the media does this model bring to the surface?*', '*Which questions about equity, sustainability and fairness does it raise?*', '*Which actions are suggested?*', '*Can this model assist students to 'see' inequity and enable them to become responsible citizens?*' Using students' bodies to represent inequity can be transformative and can lead to actions such as paying it forward.

Concluding Comments

In the preceding examples, we have attempted to document how connecting creativity with mathematics and science can be an integral way of engaging primary school students and pre-service teachers in deep learning. Other engaging primary practices utilising the visual arts include students creating crayon resist images of fire, observational sketching of birds in the schoolyard, lino printing a range of leaves, model making and environmental sculpture using found and natural material on beaches and riverbanks (Paige & Lloyd, 2002; Whitney et al., 2004). Additionally, corridors alongside teaching spaces can be adorned with co-constructed and planned mural panels with pre-service teachers and an artist-in-residence incorporating the local fauna and flora and the night sky (Paige & Whitney, 2008). Drama strategies can also involve concert performances, rolling down riverbanks when exploring incline planes and ramps, role-play using McClintock Collective experiences and assembly performances to represent curriculum being learnt (McClintock Collective, 2008).

The recent emphasis on STEM in primary (and secondary) education has uncovered challenges related to making connections between disciplines whilst maintaining the rigour of each. Creative and body-based learning strategies engage students in understanding their world in an era of environmental decline, human influence on rising temperatures, increased carbon emissions, quantity of plastics in the oceans, and species extinction. It provides a lens for engaging students in eco-justice, that is, fairness between species, fairness between generations, and fairness within generations. CBL offers a young person agency and empowerment over their world and engenders a story of hope and engagement (Garrett & McGill, 2019). Through the development of CBL approaches to mathematics and science we challenge educators to move away from worksheet-based curriculum and teacher-led (only) approaches to find alternative approaches that include the bodies, minds and feelings of their students, immersed in their local context. Such experiences stay with young people; they provide reasons for learning and connections to life-worlds that help build active and empowered future citizens.

Acknowledgements We would like to acknowledge the artist John Whitney, who constructed the drawings to help visualise the learning experiences that have been described in the text. As they say, a picture is worth a thousand words. Thank you, John.

References

Australian Academy of Science. (2020). *Primary connections: Linking science with literacy.* Canberra, ACT: AAS. Retrieved from https://primaryconnections.org.au/

Australian Association for Mathematics Teachers. (n.d.). Maths300. Retrieved from https://aamt.edu.au/teachers/resources/maths-300/

Australian Curriculum, Assessment and Reporting Authority [ACARA]. (2017). Foundation to year 10 curriculum: Science. Retrieved from https://www.australiancurriculum.edu.au/f-10-curriculum/science/

Baker, A., & Baker J. (2019). Natural maths. Retrieved from https://naturalmaths.com.au/

Ball, D. L. (2000). Bridging practices: Intertwining content and pedagogy in teaching and learning to teach: Knowing and using mathematics. *Journal of Teacher Education, 51*(3), 241–247. https://doi.org/10.1177/0022487100051003013

Berdayes, V. (2004). The body and educational practice: Some themes in the work of Paolo Freire. In V. Berdayes, L. Esposito, & J. W. Murphy (Eds.), *The body in human inquiry: Interdisciplinary explorations of embodiment* (pp. 217–231). Hampton Press.

Boal, A. (1974). *The theatre of the oppressed.* Pluto Press.

Bresler, L. (Ed.). (2004). *Knowing bodies, moving minds: Toward embodied teaching and learning.* Kluwer Academic.

Brown, L., O'Keeffe, L., & Paige, K. (2017). The colour is velvet: A transdisciplinary approach to connecting students, from a refugee background, to the natural world. *Teaching Science: THe Journal of the Australian Science Teachers Association, 63*(4), 20–32.

Callow, J., & Orlando, J. (2015). Enabling exemplary teaching: A framework of student engagement for students from low socio-economic backgrounds with implications for technology and literacy practices. *Pedagogies: An International Journal, 10*(4), 349–371.

Chartres, M. (2008). Are my students engaged in critical mathematics education? In J. F. Matos, P. Valero, & K. Yasukawa (Eds.), *Proceedings of the Fifth International Mathematics Education and Society Conference* (pp. 23–45). Lisbon, Portugal: Centro de Investigação em Educação.

Chubb, I. (2013, November 18). *Keynote address by the Chief Scientist.* Presented at the 17th National Engineering Heritage Conference, National Portrait Gallery, Canberra.

Dawson, K., & Kiger Lee, B. K. (2018). *Drama-based pedagogy: Activating learning across the curriculum.* Intellect.

Dewey, J. (1938). *Experience and education.* MacMillan.

Duatepe-Paksu, A., & Ubuz, B. (2009). Effects of drama-based geometry instruction on student achievement, attitudes, and thinking levels. *Journal of Educational Research, 102*(4), 272–286.

Fleming, M., Merrell, C., & Tymms, P. (2004). The impact of drama on pupils' language, mathematics, and attitude in two primary schools. *Research in Drama Education: THe Journal of Applied Theatre and Performance, 9*(2), 177–197.

Frankenstein, M. (1983). Critical mathematics education: An application of Paulo Freire's epistemology. *Journal of Education, 165*(4), 315–339.

Frankenstein, M. (2001). Reading the world with Math: Goals for a critical mathematical literacy curriculum. In *Mathematics shaping Australia, Proceedings of the 18th Biennial conference of the Australian association of mathematics teachers.* Canberra, ACT: AAMT.

Freire, P. (1996). *Pedagogy of the oppressed* (rev). Penguin.

Garrett, R., Dawson, K., Meiners, J., & Wrench, A. (2018). Creative and body-based learning: Redesigning pedagogies in mathematics. *Journal for Learning through the Arts, 14*(1). https://doi.org/10.21977/D914136982

Garrett, R., & McGill, B. (2019). Fostering inclusion through creative and body-based learning. *International Journal of Inclusive Education* (Advance online publication), 1–15.

Garrett, R., & Wrench, A. (2016). 'If they can say it they can write it': Inclusive pedagogies for senior secondary physical educational. *International Journal of Inclusive Education, 20*(5), 486–502.

Hardiman, M., Rinne, L., & Yarmolinskaya, J. (2014). The effects of arts integration on long-term retention of academic content. *Mind, Brain, and Education, 8*(3), 144–148.

Hirsh, R. A. (2010). Creativity: Cultural capital in the mathematics classroom. *Creative Education, 1*(3), 154–161.

Ivinson, G. (2012). The body and pedagogy: Beyond absent, moving bodies in pedagogic practice. *British Journal of Sociology of Education, 33*(4), 489–506.

Koch, S. C. (2006). Interdisciplinary embodiment approaches: Implications for creative arts therapies. In S. Koch & I. Brauninger (Eds.), *Advances in dance movement therapy: Theoretical perspective and empirical findings* (pp. 17–29). Berlin, Germany: Logos Verla.

Leonard, A. (2009, April 22). *The story of stuff project* [Streaming video]. Retrieved from https://www.youtube.com/watch?v=9GorqroigqM

Massumi, B. (2002). *Politics of affect*. Polity Press.

Master Communications. (2010, May 12). *If the world was a village of 100 people* [Streaming video]. Retrieved from https://www.youtube.com/watch?v=FtYjUv2x65g

McClintock Collective. (1988). *Getting into gear: Gender-inclusive teaching strategies in science developed by the McClintock Collective*. Canberra, ACT: Curriculum Development Centre.

Meloney, D. (2016, May 23). A body-based approach to teaching maths. *Teacher*. Retrieved from https://www.teachermagazine.com.au/article/a-body-based-approach-to-teaching-maths

Neuschwander, C. (2001). *Sir cumference and the Great Knight of Angleland.* Watertown, MA: Charlesbridge.

Nguyen, D., & Larson, J. (2015). Don't forget about the body: Exploring the curricular possibilities of embodied pedagogy. *Innovative Higher Education, 40*(4), 331–344.

O'Keeffe, L., & Paige, K. (2019). Re-highlighting the potential of critical numeracy. *Mathematics Education Research Journal (advance Online Publication).* https://doi.org/10.1007/s13394-019-00297-8

Osler, J. (2007), A guide for integrating issues of social and economic justice into mathematics curriculum. *Radical Math.* Retrieved from http://radicalmath.org/main.php?id=resourc

Paige, K., Lloyd, D., & Smith, R. (2019). *Intergenerational education for adolescents towards liveable futures.* Newcastle-upon-Tyne, UK: Cambridge Scholars.

Paige, K., & Lloyd, D. (2002). Rocks, leaves, water and bark: An environmental art experience. *Investigating Primary and Junior Primary Science Journal, 18*(2), 23.

Paige, K., & Whitney, J. (2008). Vanishing boundaries between science and art: Modelling effective middle years of schooling practice in pre-service science education. *Teaching Science, 54*(1), 42–44.

Robinson, A. H. (2013). Arts integration and the success of disadvantaged students: A research evaluation. *Arts Education Policy Review, 114*(4), 191–204.

Sherman, H. J., Richardson, L. I., & Yard, G. J. (2015). *Teaching learners who struggle with mathematics: Systematic intervention and remediation* (3rd ed.). Pearson.

Siemon, D., Bleckly, J., & Neal, D. (2012). Working with the big ideas in number and the Australian curriculum: Mathematics. In B. Atweh, M. Goos, R. Jorgensen, & D. Siemon (Eds.),*Engaging the Australian national curriculum: Mathematics—Perspectives from the field* (pp. 19–45). Wahroonga, NSW: MERGA.

Sleeter, C. E. (2012). Confronting the marginalization of culturally responsive pedagogy. *Urban Education, 47*, 562–584.

Stevens, R. (2012). The missing bodies of mathematical thinking and learning have been found. *Journal of the Learning Sciences, 21*(2), 337–346.

Van de Walle, J., Karp, K., & Bay-Williams, J. (2015). *Elementary and middle school mathematics: Teaching developmentally* (9th ed.). Pearson.
van Kraayenoord, C. E. (2007). School and classroom practices in inclusive education in Australia. *Journal of Childhood Education, 83*(6), 390–394.
Walker, E., Tabone, C., & Weltsek, G. (2011). When achievement data meet drama and arts integration. *Language Arts, 88*(5), 365–372.
Whitney, J., Sellar, S., & Paige, K. (2004). Science, art and teaching: Making connections. *Teaching Science, 50*(4), 25–29.
Yoo, J., & Loch, S. (2016). Learning bodies: What do teachers learn from embodied practice? *Issues in Educational Research, 26*(3), 528–542.
Zevenbergen, R. (2000). 'Cracking the code' of mathematics: School success as a function of linguistic, social and cultural background. In J. Boaler (Ed.), *Multiple perspectives on mathematics teaching and learning* (pp. 201–223). JAI/Ablex.

Kathryn Paige (B.Ed., Dip.T., M.Ed., Ph.D.) is an associate professor in science and mathematics education at the University of South Australia. She taught for seventeen years in primary classrooms in a range of schools, rural, inner city and in the United Kingdom, and in science teacher education for the last twenty years. Kathryn's research interests include pre-service science and mathematics education, eco-justice and place-based education. Current projects include citizen science, water literacies, connecting children to the natural world and STEM and girls. Past projects include creative body-based learning, redesigning pedagogies in the north, and distance education in the Eastern Cape, South Africa.

Leni Brown (B.Ed., B.Teach., MSS, M.Ed.) is a lecturer in mathematics and science education at the University of South Australia. Her teaching career has spanned a range of primary school settings for 25 years and she has been involved in teacher education for the past 18 years. Leni's educational interests include the use of creative body-based learning to enhance educational outcomes in mathematics and science and girls in STEM.

Lisa O'Keeffe (B.Sc., Grad. Cert., Ph.D.) is a senior lecturer in mathematics education in the School of Education at the University of South Australia. She has also worked in the Irish and UK contexts. Lisa's main research focus is mathematics textbook analysis, the role of mathematical language, numeracy and girls in STEM. Her recent work considers the role of analysis of language for mathematical learning and assessment and she was awarded the MERGA 2016 Early Career Researcher award for her paper "A preliminary analysis of the linguistic complexity of numeracy skills test items for pre-service teachers".

Robyne Garrett (B.Ed., GMD, M.Ed., Ph.D.) is a senior lecturer in physical education, dance and teaching methodologies in the School of Education at the University of South Australia. Her research interests include gender, critical pedagogies, embodiment and arts-based instructional approaches. Current projects include creative and body-based learning, socially just pedagogies for disadvantaged students, and constructions of gender in movement forms. Her research methodologies include action research, narrative storytelling and ethnographic approaches. In her teaching role she focuses on supporting student teachers to develop and implement critical and embodied pedagogies.

Responding to Climate Change: Developing Primary Children's Capability to Engage with Science Through Drama

Delia Baskerville and Dayle M. Anderson

Abstract Using drama to position children as expert scientists working on a given commission, supports them to learn about Nature of Science (NoS). Our research working with primary children uses this approach drawn from Heathcote's Mantle of the Expert (Heathcote & Bolton, 1995) to support learning about NoS. The New Zealand Curriculum (NZC) places learning about NoS as overarching and compulsory with the purpose of developing children's science capabilities for citizenship (Bull, 2007). However, children often struggle to develop and identify learning about NoS that is useful for scientific literacy (Allchin, 2014). Drama offers creative ways to consider real world issues. Our research therefore aimed to integrate drama and science in equal partnership to help children learn about NoS; we wanted to develop children's science capability while building their understanding of the causes and consequences of climate change and ways to respond. This chapter presents the refinements made when using two drama tools during the three iterations of this drama-science inquiry process, and identifies the support children need to develop capabilities and understandings about science that help them in engaging with science-related issues such as climate change.

Keywords Nature of Science · Drama · Integrated curriculum

Background

When speaking for climate justice, Thunberg (2019) claims that we are living in a time of potential catastrophic risk for humanity and non-human species. She asserts that we face chronic inequality within and between countries of the world. The underfunded health systems in poor countries with chronic poverty-related diseases impact worldwide because diseases are spread between nations through tourism; the

D. Baskerville · D. M. Anderson (✉)
Victoria University of Wellington, Wellington, New Zealand
e-mail: dayle.anderson@vuw.ac.nz

D. Baskerville
e-mail: delia.baskerville@vuw.ac.nz

P. J White et al. (eds.), *Science and Drama: Contemporary and Creative Approaches to Teaching and Learning*, https://doi.org/10.1007/978-3-030-84401-1_6

sufferings of the many pay for the luxuries of the few. Robertson (2012) justifies this further, stating that we live in mind-supremacist and outcome-driven market places that promote individualism at the expense of more collective approaches. Saying that, individuals need to wake up and change; it needs to start now; then, and only then, will hope come (Thunberg, 2018). Education is an important part of the solution, not just as a means of preparing children for the global workplace, but as an opportunity to participate in educational enterprise and focus on social justice concerns (Robertson, 2012). Our concern relates to the urgency of educating children, without alarming them, about the current real worldwide issue of climate change.

We suggest that drama, specifically the distance frame of the fictional context, provides a safe space for children to participate in such work; to take on the role of environmental scientists and explore the issue of climate change while at the same time learning about the way science works. "This distance frame provides some protection from the external consequences for those who step into it" (O'Toole, 1992, p. 26). This chapter then describes how children in-role as environmental scientists were supported to learn about NoS and develop science capability while building their understanding of the causes and consequences of climate change and ways to respond in an informed way. We present the refinements made during the three iterations of this drama-science inquiry process and identify the support children need to develop capabilities and understandings about NoS.

Pedagogical Practices of Drama

Drama is an art form: a dynamic happening within a fictional context that engages participants in activities cognitively and affectively, objectively and subjectively (O'Toole, 1992). This enjoyable, personal experience and social form of learning draws on children's curiosity, extending their innate ability to play (Drama New Zealand, 2013). Children are expected to focus on the elements of drama (focus, action, role, tension, time and space) to create imagined worlds in order to explore and express human experience (Ministry of Education, 2007). They experience opportunities to develop confidence, critical thinking skills, empathy and make connections to their world (Luton, 2014). Children are able to imagine possible futures through reflection on the past and present (Neelands, 2006). Within the drama learning space, through manaakitanga (valuing integrity, trust, sincerity and equity) children build positive relationships that affirm the identity of each child, and enhance their understanding of others and cultural perspectives (Baskerville, 2011). "Drama can change lives and give children principles they can learn to live by" (Bolton, 1986, p. 8).

Children have opportunities to learn when engaging in the two different forms of drama: (1) learning *through* drama; and (2) learning *in* drama. When learning *through* drama—a creative form of inquiry—children explore their own lives, question and engage in learning conversations (Drama New Zealand, 2013). Whilst learning *in* drama, children explore the language and forms of theatre and are expected to be in-the moment, collaborate and also contribute to formal performance. In these roles, as play-makers and actors, children experience a sense of agency and work towards positive outcomes (Drama New Zealand, 2013). Drama does not imitate life passively; drama is a way of thinking, ordering and categorizing life (Bolton, 1986).

Science, Drama and the New Zealand Context

Understanding NoS is an overarching goal of the science learning area of the New Zealand and other international curricula (Ministry of Education, 2007); historically this goal has proved difficult to achieve (e.g. Lederman & Lederman, 2014). However, drama has been shown to have the potential to support learning in three major areas important in science education: understanding science concepts, that is, the ideas and theories produced by science; considering the ways in which science and society interact; and learning about NoS, the processes through which scientific claims are developed and accepted (Ødegaard, 2003). Analogical role-play has been used successfully to develop scientific concepts (Braund, 2015). McSharry and Jones (2000) suggest ways in which role-play can offer cognitive development as well as promote procedural understanding. Toonders et al. (2016) found that a process involving script writing and performance supported science students to consider ethical implications of socio-scientific issues. Thus it appears that different drama approaches may address different learning goals in science.

Warner and Anderson (2004) positioned children as expert scientists undertaking a given commission; this role positioning supported significant improvements in children's writing and understanding of science content. We wondered about the potential of such an approach to help primary children understand something of what it means to work and behave like a scientist; that is, to learn about NoS, while at the same time learning through drama. We have designed a guided drama-science inquiry process which draws from Heathcote's Mantle of the Expert (Heathcote & Bolton, 1995): primary children are positioned as environmental scientists in a company receiving a commission. Such an approach fits well with the intentions of the NZC: to enhance the likely links between disciplines in learning experiences in classrooms (Ministry of Education, 2007). NZC places learning about NoS as overarching and compulsory. Additionally, through drama, NZC claims that children learn to enact roles and inhabit fictional worlds; it is through this purposeful play that they discover how to link imagination, thoughts, and feelings, gain new authority and examine the attitudes, actions, and beliefs of someone else. So far, we have carried out three iterations of our process, all set in the context of a fictional company of environmental scientists.

Designing and Resigning Through Iterative Cycles

One problem when integrating curriculum areas is that the goals of one discipline are often subsumed in meeting those of another. An important focus for us as science and drama educators is that the aims and purposes of both drama and science are met through the designed process. The Treaty of Waitangi (Māori: Te Tiriti o Waitangi) is New Zealand's founding document, signed on 6 February 1840. The Treaty is an agreement, written in Māori and in English, which was made between representatives from the British Crown and many Māori chiefs. Three articles from this document underpinning the relationship between the Government and Māori–often referred to as the principles of partnership, participation and protection—were the foundation on which our collaboration between science and drama was established. We formed an alliance to collaborate in developing learning about climate change and NoS through drama, and agreed to be equal partners, neither discipline becoming subservient to the other. These separate bodies of knowledge would be protected and our work would honour their unique purposes and educational intentions. We would work together to develop understanding of climate change and how to respond to it and generate principles for design (Anderson & Shattuck, 2012) for two separate disciplinary outcomes: (1) to identify the nature of learning in drama when children are preparing to adopt the role of environmental scientists; and (2) to identify what children learn about NoS during this drama-science inquiry process.

Our previous research indicates that through participation in a fictional scientific company, children readily associate with science many aspects that they have experienced in-role, such as the need for perseverance, resilience and reliable data, and for working collaboratively (Baskerville & Anderson, 2015). We also noted that insufficient scaffolding and support impacted on the level of understanding of scientists' work that was demonstrated by children when in-role. Anderson (2004) suggested that an 'as if' laboratory must provide learners with sufficient scaffolding to engage in meaningful roles. Other analyses of our work (Anderson & Baskerville, forthcoming) have examined across iterations the implementation of drama tools including Hot-Seating and Mime-Freeze with side-coaching, both using teacher-in-role. The findings align with socio-cultural theoretical framings of learning in apprenticeship situations (e.g. Lave & Wenger, 1991). Students incorporated more scientific ways of working into their role-taking when they had opportunities to hear about scientific practice and then enact it. Another effective scaffold was the use of side coaching that prompted scientific practices as students were working in-role, and simulated work situations where novices learn from a more expert practitioner who models, coaches and supports the adoption of the accepted practices and values of the community.

In this chapter, we focus on two other drama tools used in each iteration of the process, the 'Staff Notice Board' (SNB) and 'Teacher Role Play' (TRP). We adapted these in different iterations, with the intention of making NoS more explicit; supporting children to develop authenticity and disciplinary language and knowledge to prepare them to enter the role of environmental scientists. For instance, we used

TRP to explicitly model science practices such as making and critiquing claims, or careful observation and evidence-based inferencing. Other adaptations were less experiential, such as collaborative close reading of information, used in conjunction with the SNB. Here we consider children's responses to our adaptations in iteration three to then consider the kind of support children need in drama to develop capability and understanding about science that is useful for mitigating climate change.

Methodology

Our project aimed to investigate how entering a fictional science world in the classroom supported children to understand NoS while also meeting curriculum goals for drama. This paper explores two questions: (1) What scaffolding provides children with an opportunity to build accurate language and knowledge to inform role-taking as an environmental scientist?; and (2) How does preparation for a fictional science world impact on their ability to contribute to meeting expected outcomes for both drama and science?

We have completed three iterations of the process, each with three different classrooms of children in two different primary schools using a design-based research (DBR) approach. DBR aims to find solutions to a problem within complex contexts like classrooms. Researchers and teachers work together, using knowledge of context and research evidence to design and refine potential solutions through iterative cycles to provide practical guidance concerning the problem (Anderson & Shattuck, 2012). Reflection on iterations one and two identified the problem that children required further scaffolding to firstly, identify scientists' practices and values and secondly, to develop specific science capabilities such as making and critiquing claims. This prompted us to refine our use of two drama tools: the staff notice board and teacher role-play.

All three iterations were set in the context of a fictional company of environmental scientists and aimed to support children to identify personal and community actions to mitigate effects of climate change. Each had a different commission specific to and informed by the class context. The first involved a commission where children designed a fictional 'low-emissions' race track; in the second, children were commissioned to prepare an exhibition informing local residents about global warming; and in the third, the commission was to identify the impact on a fictional town of a local consequence of global warming, increased frequency of intense rainfall, and to suggest appropriate mitigation. All three iterations introduced dramatic tension when a high-status person, who represented the commissioning body, appeared in-role and announced a threat to their commission due to inadequate funding. Children's successful justification (and verification of learning) caused the commission representative to reconsider the allocation of funding. Participants in each iteration were different and comprised a class of about 25 middle primary children (aged 8–11). Data informing this chapter included transcripts of audio-recordings of classroom activities and student work. Data sets were analysed inductively, emerging codes were identified and compared across the three iterations of the inquiry process.

Findings

We now present our findings from the third iteration of the process, to identify support that children needed to develop science capability and understanding about science useful for mitigating climate change. We focus on two drama tools: a drama prop (SNB) and a drama convention (TRP). Innovations made as a result of the two previous iterations were designed to support children to develop disciplinary knowledge and disciplinary practice.

Staff Notice Board (SNB)

The SNB, a prop used in all iterations of the inquiry processes, supported children to experience and practice authentic social behaviours of scientists in a safe setting. This prop was designed to prepare children to enter the role of a scientist, building belief in a company they would name and be part of to consider a real-world issue: climate change. The large SNB prop was displayed on the wall as children re-entered the room after interval. There were several sections: work news, with pictures of science teams undertaking fieldwork; staff news with photos of people engaged in social and sporting events; information on upcoming conferences encouraging staff to attend and share their work; recommendations from clients describing work done previously; staff excellence awards listing what they had done well in their work; scientific information about climate change and a notice about the importance of staff keeping up to date with what other scientists were doing in this area. Items were created to provide details and information we perceived as relevant to children developing understanding about NoS. During the first process, children were encouraged to look at the SNB to find out about the company and the people who work for it. Most only noticed personal information related to their own lives, for example their personal names and birth dates. A similar outcome occurred in the second iteration, students noticed personal information about the scientists, but not so much about the nature of the company business or its values.

The innovation examined here provided structured questions prior to children engaging with the SNB to support them to look for specific information related to NoS. Children were scaffolded in two ways. Firstly, they were asked to choose a partner and observe and gather information in order to answer their allocated, specific question. Secondly, they were given an opportunity to share the information they had gathered with the class. There were six questions:

1. Who are these people who work in this space?

2. What work is important to this company?
3. What does this company do well?
4. How does this company treat its people?
5. What did you notice about the messages that it gives this company?
6. What ideas, values and beliefs are important to this company?

These questions supported children to record relevant ideas Earlier in the process, data was gathered utilising the role-on-the-wall convention, which served as a diagnostic test to report children's prior knowledge of scientists; children identified stereotypical perceptions of scientists in this activity. When reading the SNB they recorded notes about scientists as people with identified names and the importance of working in teams. Children commented on aspects related to scientists collecting samples, using instruments on location and that their work required precision. They also identified the purpose and high-quality of their work, noted the ethical nature of the company, and perceived the company as valuing equity. One noteworthy aspect of children's observations was children positioning themselves in the company by using the words 'our company', indicating that they were building belief in the fictional context. However, children tended to copy the information directly from the SNB and although they identified relevant ideas to answer their question, they often did not interpret the information at this point.

A second form of scaffolding followed up to the SNB activity—a post-observation reflection time. Individual children presented ideas they had gathered from the SNB in their work groups to the rest of the class. Again, this discussion was focused in relation to the given questions (listed above). When presenting their ideas to the class, children provided evidence of their developing body of science knowledge. Children reported that environmental scientists carry out work in the field, collect data, help the environment, and are aware of what other scientists are working on. Children perceived scientists' as research active; they gave examples of the special events that scientists in this company participated in. In this activity, children were more likely to use their own words to describe their ideas. This post-observation group reporting time provided an opportunity for children to decode the noticeboard prop and indicated they were able to identify some of the things scientists do and how they work.

The SNB supported children to read and identify accurate information about scientists; they were beginning to develop a drama body of knowledge. Children were able to see environmental scientists as humans; perceive them as travellers in time and space; and consider the development of an ethical company in a fictional world. They appreciated the role of management in creating positive company relationships, which could inform their co-construction of this company. Children positioned themselves in this fictional company. This SNB supported children to develop their understanding of environmental scientists, prepare for their role-taking and built belief in this fictional world. Although children participated in this part of the learning process and could develop ideas based on accurate knowledge about scientists, they needed further opportunities to process the information. As ideas that are enacted seem to support children to further develop capabilities and understandings about

NoS (Baskerville & Anderson, 2015), this limitation requires further opportunities for children to be in-role and play as scientists in order to process information in future iterations.

Following this activity, we invited children to talk with their partner about an item they would need if they were to work in this company. In this discussion, children used their powers of invention to identify an object that they would add to the noticeboard and enhance the creation of the dramatic world. This whole class activity engaged children (one group at a time) to explain their pictures for the staff noticeboard. These were mainly individual ideas: Laptop, 8/30; chocolate, 3/30; water, 2/30; hot chocolate, 1/30; white light, 1/30; plane, 1/30; whiteboard, 1/30; wifi 1/30; table, 1/30; dog, 1/30; camera, 1/30; coffee, 1/30; food, 1/30; iPod, 1/30; iPhone, 1/30; music, 1/30; fruit, 1/30; strawberry, 1/30.

Adding this object to the noticeboard was one of the factors that made it possible for children to begin to invest in this company and begin to build belief in this fictional world. O'Neill (1995) supports this claim: the props that children select "assist them in creating their world of illusion" (p. 160).

A follow up 'think pair share' activity also provided scaffolding for children to contribute ideas to the class in their groups. Children were asked what they might need to work with as scientists in this company. Some children reported ideas that teachers mentioned previously, "What evidence do I need to look for?". Another child built on the ideas s/he read on the noticeboard, "How will I test my evidence?" Whilst other children reported evidence of original thinking, "What tools will I need?", "What will I use?", "How is this going to help the environment?", "What am I doing?", and "You could use gloves". These children wondered about what would help them individually to do the job and their possible approach to the work in order to problem-solve. The contributed ideas indicated that some children were beginning to understand aspects about NoS: that scientists' problem-solve and gather evidence.

Teacher Role-Play (TRP)

Role-playing is defined as "using the imagination to identify with someone else in order to explore and represent experience from their perspective or viewpoint; also called being in-role" (Ministry of Education, 2000, p. 49). The TRP convention was introduced during the first iteration of our process because we had not sufficiently prepared children to critique the claims of other scientists. The TRP convention introduced aspects that are an important part of the work that is undertaken by expert environmental scientists. There were three purposes for introducing this convention at this time: (1) to introduce and model a conversation between experts that contextualises learning; (2) to develop accurate knowledge; and (3) to model how experts use their skills, knowledge and understanding in action. The TRP modelled scientists' behaviours: making detailed and accurate observations and forming inferences based on observations. Two teachers in-role made observations and inferences about some photographs, also visible to the children, related to rainfall and erosion. During the

TRP, children made notes in their groups related to given questions: 'What do you see in the photograph?', 'What do you think the roots will do to the soil?', 'What sense did Dayle make of the photograph?', 'What did [teacher name] think about it?', 'And what are the implications for your area of the town when there is too much rain?' They were able to note observations (big roots, washed away dirt, not much grass, lots of twigs), separating these from inferences (the roots will rip the ground apart and make the tree fall down) and also made inferences on observations (if there is too much rain, trees will fall over). The TRP supported children to develop their capabilities to work in scientific ways. The guided questions and note-taking were effective strategies that assisted children to achieve this.

The TRP successfully modelled authentic language so children were able to describe disciplinary specific ideas about the nature of scientists' work and apply some aspects of scientists' skills, knowledge and language in their work. This imaginary, scripted TRP narrative provided an opportunity for children to witness authentic clarification and feedback between scientists that they transferred into their own practice of science during the activities that followed. The children went on to work with sets of photographs of different landscapes present in the fictional town and make inferences based on their observations about what might happen during intense rainfall. They also carried out practical activities in groups investigating the effect of water on different parts of the school grounds, for example paved surfaces with and without gutters and drains, loose soil, planted and bare banks.

All groups were able to make detailed observations and most made sound inferences from them. For example, using a photograph of an industrial area, one group observed workshops, buildings, roads, cars, trucks, flat paved areas, and grates. They inferred that during heavy rainfall "the water might stay or go down the grate"; "water will enter the buildings". For a similar image without obvious drains their inference was "there doesn't appear to be a way for water to leave". They recommended that the council provide "lots of grates", "put more drainage in" and "wide guttering". Students also drew on observations from their practical investigations in making recommendations to the council about preparing for climate change. One group investigating water poured on a steep slope, observed that "the ground under grass absorbed water and ran down hill, but when someone stood on the ground the grass ripped leaving only soil." Their recommendations about eroded grassy slopes appeared to draw on their practical observations: "don't put anymore animals on it because it could course (sic) more erosion", and "let paddocks regenerate so that the grass will grow back". Another group that observed leaves impeding the flow of water into drains outside the school, recommended to the councillor that the drains needed to be made larger and kept clear. These findings suggest that students were able to enact the scientific practices modelled during the TRP; they made observations and used them to make inferences and develop pragmatic recommendations about ways to prepare for and adapt to climate change.

Previous research (Baskerville & Anderson, 2015) indicates that children in-role as scientists work collaboratively. This chapter provides detailed evidence: children developed capability to work together in teams, listened to each other and shared ideas about what it meant to be a good scientist. They identified four features relating to

NoS. Firstly, two groups identified an aspect about the nature of scientists work: "work as a team". Secondly, children reported an attribute of effective work in a team of scientists: "listen to each other's ideas", "let everyone have a turn," and "no one dominating in a group means no one tries to take over". Thirdly, children reported on the values scientists show when working in a team: "always using kind words", "team work and encourage your whole team", "help each other", "respect each other", and "you respect them for who they are and not for who you want them to be". Finally, children offered ideas about the reality of the work of good scientists: "so you work together to get whatever you need done", and "don't give up and use your imagination". These children realised that scientists work in teams to get work done. They perceived encouragement, support, respect, listening and equal partnership as attributes of an effective member in a team of scientists. They identified perseverance and imagination as characteristics that scientists need to function in their workplace. The TRP supported them to begin to understand the ethical nature of team work and how they needed to behave in order to make a positive contribution to their work with other scientists. This was important learning through drama given the collaborative problem-solving skills children need to understand and adapt to climate change.

Children's responses to the SNB and TRP and the follow up practical activities indicate increased inclusion of accurate scientific language and behaviours, and their ability to be prepared to enter the role of an environmental scientist. In this learning process, although focused on their individual perceptions, this company was becoming their company. Positioning themselves in this way, these children were ready to consider a real-world issue: climate change. They were also beginning to invest in their future lives as environmental scientists in their company and its creation, as the responsibility of the whole class. This highlights the significance of drama in the education of values and ethics. Drama not only addressed learning goals in science, but provided the context for this to happen and for children to gain insights into scientists' ethical behaviours.

Discussion

A SNB and a TRP provided the context for this discussion. We identified these as a prop and a convention whose functions differed. The information on the SNB prop was designed so that children's guided reading built their ideas to use accurate knowledge in the fictional world. The questions we introduced provided the necessary support for children to find and report words and ideas they understood. Children simultaneously decoded the SNB and gave meaning by thinking about an ethical company of environmental scientists and how these people worked. In contrast, the TRP was a model of practice that showed them what scientists do when *being* scientists. The TRP conversation between two scientists assisted children to fulfil the two important components of scientists' work: to understand how to observe carefully, separating observation from inference: and to make inferences based on observations (Lederman & Lederman, 2014). They were developing a state of being learners, gathering information that was incorporating personal, social and collective knowledge for action; which according to Munday et al. (2015), means they had agency in their learning.

Drama and science education's unique purposes and intentions were achieved through the use of these two drama ways of working. Separate bodies of knowledge were not only protected but both developed through common pedagogical approaches: observation, description, thinking time, decoding, and imagining. Drama prepared children to enter their roles; improved their ability to think as scientists; and supported them to build belief in the fictional world. Drama was a motivating factor that provided children with the opportunity to perceive themselves as scientists in an authentic company and the agency to understand the world of scientists. Science provided accurate knowledge to equip children to make careful observations and inferences based on data. Questioning, modelling through TRP, thinking time, and pair and group discussions provided the scaffolding to meet these outcomes.

The primary function of drama education is to create imagined worlds to explore and express human experience (Ministry of Education, 2007). These two drama ways of working positioned children in a place of readiness and willingness for high-status role-taking as expert scientists in a fictional context, as part of which they were able to explore and express the experience of scientists. Being active agents in the learning process highly motivates and deeply engages children to take on challenges (Williams, 2017). This is an important step toward enacting and rehearsing scientific practices.

Conclusion and Implications

Intentionally considered scaffolding supported children to make connections between the real and the fictional world, and develop authenticity in preparation for role-taking as scientists. Informed design that requires expertise, collaboration, and respect for both disciplines, appears important in the integration of science with drama if curriculum aims for both disciplines are to be met. Further research is needed to identify how and why children select the ideas that they notice and use in-role, and the nature of further scaffolding required to help children interpret the identified information.

References

Allchin, D. (2014). From science studies to scientific literacy: A view from the classroom. *Science & Education, 23*(9), 1911–1932.

Anderson, C. (2004). Learning in 'as-if' worlds: Cognition in drama in education. *Theory into Practice, 43*(4), 281–286.

Anderson, D., & Baskerville, D. (Forthcoming). Integrating science and drama to support learning about the Nature of Science in New Zealand primary classrooms. In D. McGregor (Ed.), *Learning*

science through drama: Exploring a range of international research perspectives. European Science Education Research Association.
Anderson, T., & Shattuck, J. (2012). Design-based research: A decade of progress in education research? *Educational Researcher, 41*(1), 16–25.
Baskerville, D. (2011). Developing cohesion and building positive relationships through story telling in a culturally diverse New Zealand classroom. *Teaching and Teacher Education, 27*(1), 107–115.
Baskerville, D., & Anderson, D. (2015). Investing in the pretend: A drama inquiry process to support learning about the nature of science. *Research Information for Teachers, 1*, 50–57.
Bolton, G. (1986). Emotion in drama. In D. Davis & C. Lawrence (Eds.), *Selected writings* (pp. 164–179). Longman.
Braund, M. (2015). Drama and learning science: An empty space? *British Educational Research Journal, 41*, 102–121.
Bull, A. (2015). *Capabilities for living and life-long learning: What's science got to do with it?* Retrieved from http://www.nzcer.org.nz/research/publications/capabilities-living-and-lifelong-learning-whats-science-got-do-it
Drama New Zealand. (2013). *Drama education in the primary classroom*. Retrieved from www.drama.og.nz
Heathcote, D., & Bolton, G. (1995). *Drama for learning: Dorothy Heathcote's mantle of the expert approach to education*. Heinemann.
Lave, J., & Wenger, E. (1991). *Situated learning: Legitimate peripheral participation*. Cambridge University Press.
Lederman, N. G., & Lederman, J. S. (2014). Research on teaching and learning of nature of science. In N. G. Lederman & S. K. Abell (Eds.), *Handbook of research on science education (Volume II)* (pp. 600–620). Routledge.
Luton, J. (2014). Finding the golden ticket: Reflections on the development of drama education in New Zealand. *Pacific-Asian Education, 26*(2), 5–20.
McSharry, G., & Jones, S. (2000). Role-play in science teaching and learning. *School Science Review, 82*(298), 73–82.
Ministry of Education. (2000). *The arts in the New Zealand Curriculum*. Wellington, New Zealand: Learning Media.
Ministry of Education. (2007). *The New Zealand Curriculum*. Wellington, New Zealand: Learning Media. Retrieved from https://nzcurriculum.tki.org.nz/The-New-Zealand-Curriculum
Munday, C., Anderson, M., & Gibson, R. (2015). Drama in schools: Agency, community and the classroom. In K. Freebody & M. Finneran (Eds.), *Drama and social justice: Theory, research and practice in international contexts* (pp. 75–88). Routledge.
Neelands, J. (2006). Re-imagining the reflective practitioner: Towards a philosophy of critical praxis. In J. Ackroyd (Ed.), *Research methodologies for drama education* (pp. 15–40). Sterling, UK: Trentham.
Ødegaard, M. (2003). Dramatic science. A critical review of drama in science education. *Studies in Science Education, 39*(1), 75–101.
O'Neill, C. (1995). *Drama worlds*. Heinemann.
O'Toole, J. (1992). *Process of drama: Negotiating art and meaning*. Routledge.
Robertson, S. K. (2012). *Educating the heart and mind* [Streaming video]. Dalai Lama Center for Peace and Education. Retrieved from https://www.youtube.com/watch?v=I1A4OGiVK30
Thunberg, G. (2018). *Full speech at UN climate change COP24 conference* [Streaming video]. Retrieved from https://www.youtube.com/watch?v=VFkQSGyeCWg
Thunberg, G. (2019). *Full speech at UN climate action summit* [Streaming video]. Retrieved from https://www.youtube.com/watch?v=KAJsdgTPJpU
Toonders, W., Verhoeff, R. P., & Zwart, H. (2016). Performing the future. *Science & Education, 25*(7–8), 869–895.
Warner, C. D., & Andersen, C. (2004). "Snails are science": Creating context for science inquiry and writing through process drama. *Youth Theatre Journal, 18*(1), 68–86.
Williams, P. (2017). Student agency or powerful learning. *Knowledge Quest, 45*(4), 9–15.

Delia Baskerville (ATCL (Speech and Drama), Dip. Teaching, B.Ed. M.Ed. Drama (Hons), Ph.D.) is a pracademic; a senior lecturer in initial teacher education. She is a trained primary teacher, an actress, and an experienced secondary school drama teacher. In 2003, Delia became a teacher educator and Ministry of Education Advisor implementing the drama discipline of the Arts Curriculum in secondary schools in the Wellington region. Her on-going research—the *Mattering Project*—involves a community response to *Mattering*—an ethno-dramatic political theatre piece, which disseminated the results of her Ph.D. Community members were challenged to listen to the perspective of marginalised young people who truant and respond in post-performance discussions (talanoa).

Dayle M. Anderson (B.Sc. (Hons), Ph.D.) is a senior lecturer in initial teacher education. She worked for a number of years in science as a biochemist, later training as a primary teacher, before moving into science education research and tertiary level teaching. Her research is largely classroom based and she is involved in a number of collaborations investigating aspects of teacher practice and student learning in primary science. In addition to her research and work in initial teacher education, Dayle is involved in teacher professional learning in science for the Royal Society of New Zealand-Te Apārangi, and in primary science resource development.

New Eden: Mediating Pre-service Teachers' Conceptions of Education for Sustainable Development

Debbie Myers

Abstract In this chapter the author examines the use of Mantle of the Expert (Heathcote & Bolton, 1995) as a model of integrated curriculum design for pre-service teachers (PSTs), so they can deliver pedagogical approaches enabling their pupils to think critically about the impact of their lifestyles on the environment. Drawing on the pedagogy of dramatic inquiry (Heathcote & Bolton, 1995), primary teacher educators positioned PST's as a range of experts, working for the fictitious 'Waste Not Want Not Company'. In these roles, PSTs gained insights concerning the interplay between environmental, economic and social aspects of sustainable development. As naturalists they mapped habitats within a forest setting; as recreations officers they identified possible social and educational uses of the forest; as scientists they analysed water quality; as planners they considered the development of a settlement New Eden, and as agentic environmental advocates they prepared delegate materials for a World Conference on Education for Sustainable Development. Using these approaches, teacher educators simultaneously provided PST's with a model of integrated curriculum design for subsequent emulation on their school-based placements.

Keywords Dramatic inquiry · Education for Sustainable Development · Mantle of the Expert

Introduction

It is thirty years since a definition of sustainable development was agreed by the United Nations member states:

> Sustainable development is development that meets the needs of the present without compromising the ability of future generations to meet their own needs. (Brundtland, 1987)

In 2015 UN member states adopted the 2030 Agenda for Sustainable Development consisting of 17 Sustainable Development Goals (SDGs) (DESA, 2010). The new

D. Myers (✉)
University of Northumbria, Newcastle, UK
e-mail: debbie.myers@northumbria.ac.uk

P. J White et al. (eds.), *Science and Drama: Contemporary and Creative Approaches to Teaching and Learning*, https://doi.org/10.1007/978-3-030-84401-1_7

framework, Education for Sustainable Development (ESD) for 2030 will focus on ESD as a key instrument for the delivery of the SDGs through a Global Action Plan (GAP) for ESD (UNESCO, 2014b) However, Qablan (2018, p.130) observes:

> …some educators attribute current unsustainable practices and lifestyles to a specific deficiency in education systems worldwide-a lack of focus on helping learners to think critically about their own lifestyles.

Teacher educators have a key role to play in developing and modelling to pre-service teachers (PSTs) examples of curriculum-making that will enable their future pupils to think critically about their life-style choices (Mulà et al., 2017). In this chapter the author explains how teacher educators integrated dramatic inquiry, expeditionary fieldwork, story-telling and scientific enquiries to mediate PSTs conceptualisations of sustainable development. This example of practice illustrates participatory teaching and learning methods that support collective decision-making in accordance with the GAP (UNESCO, 2014a; b).

Dramatic Inquiry

The use of drama as a pedagogical strategy facilitates 'pedagogical border crossings', enabling teacher educators to model to PSTs how to connect subject material drawn from different disciplines (Fels & Meyer, 1997, cited in Braund, 2015, p. 112). Combining drama and science can support the examination of socio-scientific issues with PSTs including ESD (Braund, 2015; McNaughton, 2006) "by creating a learning situation that is significant in the lives of students" (Ødegaard, 2003, p. 76). Drawing upon Sjoberg's (1998) delineation of three dimensions of science as a product, process and as a social institution in society Ødegaard (2003) observes:

> Drama can successfully be used for making simulations of the real everyday world, especially in learning about science in the context of society, or where science is recontextualised for specific societal purposes. (Ødegaard, 2003, p. 97)

Dorion (2009) defines drama in science education to be a form of "role-play within an imagined situation and enacted within the human dimension" (p. 2249). Mantle of the Expert (MoE) (Heathcote & Bolton, 1995) is a form of dramatic inquiry that integrates drama for learning, inquiry-based learning and expert framing. When using MoE as a pedagogical strategy, educators position groups of learners in-role as members of a fictional company or team, who collectively undertake a range of tasks to gain deeper insights. Educators position themselves as co-investigators, non-expert agents, who act in-role on the instructions of a fictitious client who has issued the team/company with a special commission. Expert framing and fictional positioning support improvisational role-play by providing learners with a structural scaffold to simultaneously maintain these dialogues (Dorion, 2009; Ødegaard, 2003).

Using Dramatic Inquiry to Teach About Sustainability

During the development of an undergraduate module of primary teacher education, tutors specialising in English, mathematics, science and ICT identified an opportunity to use drama to provide a model of integrated curriculum-making. Science tutors positioned PSTs in-role as experts working for the fictional company 'The Waste Not Want Not Company Ltd' to mediate their conceptualisations of sustainability. They presented this company with a plausible commission on behalf of a client, the organising committee of the United Nations Educational and Scientific Organisation (UNESCO) World Conference on Sustainable Development (UNESCO).This commission would require experts from 'The Waste Not Want Not Company Ltd' to consider different options for re-development of a forest setting and to represent these views at the World Conference. Situating this dramatic inquiry in the natural setting of a forest and then in the science teaching rooms would enable the creation of a real-world scenario providing a degree of authenticity to the activities (Budzinsky, 1995) (Table 1).

This example of MoE is an adaptation of its usual format because tutors positioned PSTs in multiple rather than one, fixed expert role. Each PST therefore had an opportunity to act in-role as a naturalist, environmental scientist, developer/protester and recreation officer. The purpose of multiple framing was to give PSTs opportunities to develop insights into the different and competing motivations of environmental, economic and social interest groups, and therefore be better informed to advocate for a particular position (Table 2).

Teaching Program

The module was delivered via the following teaching sequence (Table 3).

Table 1 The Waste Not Want Not Company Ltd. Commission

Who and What	Comments
Expert team	The Waste Not Want Not Company Ltd.
Client	The United Nations Education and Scientific Organisation (UNESCO) World Conference on Sustainable Development
Commission	To investigate the current management of a forest setting with respect to environmental protection and future economic growth through the creation of a new town with homes, schools, supermarkets creating employment, recreation and leisure facilities. To prepare delegates' presentations for delivery at the conference to represent different perspectives including those of the diversity of plants and animals of the forest who have no voice
Theme	To understand that sustainable development requires a balance between the interplay of three competing pillars: environmental, economic and socio-cultural development

Table 2 Dramatic roles, fictional identities and positionality

Fictional identities	Positionality	Curricular inquiries	Location
Naturalist	Expert	Science: exploration, habitat mapping, identifying, classifying species of plants and animals	Forest setting
Environmental scientist	Expert	Science investigations: measuring water quality and pH, separating plastics, analysing leaf litter	On campus science room
Developer/protester	Expert	English: debating the tensions between environmental, economic and socio-cultural developments	Forest then on-campus
Documentary maker	Expert	Computing/ICT: building a persuasive case to support an argument for or against the development of the forest setting	Forest setting
Recreation officer	Expert	Mathematics/exploration: identifying the potential of the forest setting to support educational, recreational and leisure trails and activities	Forest setting
Agentic advocate	Expert	Arguing a position for sustainable development that balances competing demands: environmental, economic and socio-cultural	On-campus

Introductory Conference

All PST's and tutors attended a whole day conference entitled: 'Student Enquiry: Creativity and the Core Curriculum'. The conference comprised of keynote lectures, practical workshops on English, mathematics, science and ICT, and practitioner-led workshops outlining creative approaches to teaching and learning including the use of drama. At this conference, the module aims, expected learning outcomes, timetable and assessments were shared with PSTs. The keynote lectures focused on creativity, its importance to future economic development in the United Kingdom, and on the implications of neuroscience for the development of teaching and learning. The pedagogical strategy MoE was introduced and PSTs attended a briefing about the visit to the forest setting. By the end of the conference day PSTs understood the purpose

Table 3 The teaching and learning sequence

PART 1: Creativity conference	PART 2: Fieldwork in a forest setting: expert framing	PART 3: On-campus workshops: expert framing	PART 4	PART 5: Assessments
Keynote lectures and workshops	English, (economic development), Mathematics, (social development) Science, (environmental development) ICT, (defending a position).	English, Mathematics, Science, ICT	Teaching placements	Group presentations
		Science 1: Investigations		Essay
		Science 2: Developing a setting with respect to three pillars of sustainable development		Module evaluations
		Science 3: Planning teaching, learning and assessments for placement		

of the module was to model creative pedagogical design and curriculum-making for emulation by PSTs on placements and in employment.

PSTs were invited to take part in a research study to evaluate the effectiveness of these approaches in supporting PSTs development of creative and integrated pedagogical design in readiness for placement and future employment (Myers, 2019). Underpinned by an interpretivist paradigm a case-study was developed using mixed methods of data collection to elicit personal and contextual foci for analysis. Data was collected in the form of pre- and post-teaching questionnaires and focus group discussions. In accordance with British Educational Research Association (BERA) guidelines an ethics checklist was prepared and ethical approval was sought from the employing university to carry out the study to inform a Masters dissertation research study. The purpose of research was shared with those students who expressed interest in participating. Information and consent forms were provided together with explanations of how data would be collected, stored and used. Assurances were given that anonymity and confidentiality of participants would be maintained throughout the study. An extract of a focus group discussion relevant to the use of dramatic inquiry is shared in this chapter.

Fulfilling the Commission

Expeditionary Fieldwork in a Forest Setting

The forest is a remnant of a medieval hunting forest now covering 2400 acres. Little of the ancient woodland now remains due to developments over many centuries and it now contains a mixture of deciduous and evergreen trees. The forest is located over one hour from the campus so PSTs travelled there by car, coach and rail for a day of workshops. All PSTs participated in-role in four workshops in English, mathematics, science and ICT during the day.

Science Expeditionary Workshop: Mapping Habitats

Science tutors positioned PSTs in the expert role as naturalists in the forest. Following in the footsteps of Charles Darwin, Joseph Banks and Mary Anning, PSTs undertook an expeditionary "journey in pursuit of knowledge" (Hodacs, 2011, p. 187) in the forest setting to replicate the habitat mapping expeditions of these early naturalists. PSTs sketched the natural features of the setting and documented evidence of human activities (eg litter, signage, roads, fencing, lighting, first aid boxes). Working in small groups, PSTs carried out habitat mapping activities in areas of the forest. These included exploring and documenting the occurrence and diversity of plants and animals at one metre transects through the forest, using quadrats and identification charts of British flora and fauna (http://www.naturedetectives.org). While mapping transects from sunlit to shady areas PST's discovered newts-a protected species-and were overheard role-playing this discovery in the voice of naturalist David Attenborough "and so we find… hiding amongst the damp undergrowth… a rare species, a protected species- the increasingly rare, smooth newt" (Fig. 1).

Fig. 1 In role as naturalists, PSTs discover a smooth newt during habitat mapping activities

PSTs were transfixed and impressed to observe how camouflage had effectively concealed the presence of the newt. A search for newts was carefully undertaken, resulting in squeals of delight at each new find.

English Workshop: Adopting a Position

Tutors-in-role presented PSTs with a proposal to develop a substantial part of the forest setting including the area of special scientific interest that provided habitat for a protected species (the newts). This development would create a new eco-housing estate, two primary schools, a secondary school, science business park, shopping centre and recreation sites to create employment and investment opportunities. PSTs were placed at random in two equal groups with the purpose of adopting and defending a position on the issue. Sustainable Development Goal 15 requires member states to:

> Protect, restore and promote sustainable use of terrestrial ecosystems, sustainably manage forests, combat desertification, and halt and reverse land degradation and halt biodiversity loss. SDG 15 (UNESCO, 2014b).

Mathematics Workshop: Identifying Recreation and Learning Opportunities

A mathematics tutor positioned PSTs as recreation officers to consider the development of the setting for families and a variety of user groups, for example, cubs, scouts, Brownies, girl guides, forest school, walkers, nature lovers. PSTs mapped a potential range of trails to support learning, including treasure hunts, trails to look for shapes, patterns, counting, measuring, to collect natural objects and Green Man investigations using a range of trees.

ICT Workshop: Documentary-Makers Defending a Position

An ICT tutor positioned PSTs as documentary makers defending the position they adopted in the English workshop so that they developed sympathy towards one pillar of sustainability and were prepared to advocate on behalf of particular interests: environmental (protester), economic (developer), social (mediation between environmental and economic needs). PSTs created short documentaries to support arguments for or against the development of the setting. These documentaries would be edited and shared during ICT sessions on-campus.

Fulfilling the Commission: On-Campus Inquiries

Story-Telling and Science Investigations

PSTs were positioned as scientists to resolve damage resulting from human consumption and productivity in the woodland setting. Tutors shared the allegorical picture-book *The Tin Forest* (Ward & Anderson, 2001) with PSTs to contextualise their scientific inquiries. Rich in symbolism, *The Tin Forest* is an allegorical tale in which the deployment of visual metaphors allows readers to observe the damaging effects of unsustainable consumption and production on the environment due to their disconnection from nature (Louv, 2005). The main character however, personifies environmental agency and the potential for change. As he takes responsibility for initiating the positive changes he wishes to see in the environment and restores a living eco-system in a scrap metal tin forest, he enables the nourishment of his psyche and reconnection with the dynamic and healing energies of nature. PSTs discussed the themes of the story, and then worked in groups to complete a carousel of science tasks.

Enquiry 1: Water Quality Analysis

PSTs analysed the quality of water samples collected from streams in the forest and on-campus, using the OPAL Citizen Science Survey resources (Imperial College London, n.d.) (Fig. 2).

PSTs measured the turbidity and pH of the samples: water from the stream in the forest was found to be pH 7, as expected to support healthy plant and animal life. The OPAL survey (Imperial College London, n.d) showed the forest water to be of high quality because it was found to contain macro-invertebrates and therefore to support life. However, the sample from campus was of poor quality and contained no visible organisms or plants. The difference in the quality of water samples created opportunities for discussion on the possible negative impacts of human activities on environmental resources e.g. water, land and the presence of life forms. One student had noticed empty beer cans in this stream, which ran parallel to new student accommodation, and raised the issue of possible pollutants in the river e.g. alcohol. Another queried the possible leakage of building materials from the site when the accommodation was under construction.

Enquiry 2: Nature's Recycling Mechanisms

PSTs examined three samples of leaf litter for evidence of decomposition and then investigated the drainage of the samples. This activity provided an example of how nature recycles materials through processes of decomposition that contribute nutrients to the soil to support the life-cycles of plants and sustain eco-systems.

Fig. 2 In-role as scientists PST's analyse the quality of water samples from the forest and on-campus

Enquiry 3: Sorting Materials

A selection of litter generated from packaging was collected from campus. It was sorted into paper/card, plastics and metals, and examined to see how these items could be recycled, up-cycled or re-used.

Enquiry 4: Recycling Paper

Pre-shredded wastepaper was soaked in water to create a pulp. Food colouring, leaves, petals and seeds were added and the pulp was poured onto a mould screen (a frame with a fine mesh). A deckle (resembles a picture frame) was placed on top to hold the pulp in place and create the edges of the paper. Recycled paper was produced for future use as wrapping paper, textured writing paper, carrier bags, papier-mâché containers and concertina folders.

In-Role as Developers

Drawing on the collective learning generated during the range of workshops in the forest, PSTs were tasked to discuss how they could develop a substantial part of the forest setting including the area of special scientific interest (newt habitat). This re-development would create a new housing estate, two primary schools, a secondary school, science business park, shopping centre and recreation sites to create employment and investment opportunities and to provide low-cost eco-housing. PSTs were tasked to be mindful of the following inquiry questions to guide these discussions:

1. What is more important: continual economic growth, providing homes and jobs for people, or protecting the environment and its ecosystems?
2. How do we ensure sustainable development and achieve balance between competing environmental, socio-cultural or economic needs?

PSTs agreed the need to provide secure housing for people was a priority and that the development of the forest setting was an opportunity to create a model of good practice to ensure protection of the environment through responsible stewardship:

Student 1: If you look at the letters in Need and you re-arrange them, they spell out Eden-as in Garden of Eden. It's a bit like what happened in The Tin forest-the old man created a whole eco-system from junk and it became a real eco-system. Well couldn't we, as the developers, couldn't we do the same. Couldn't we link the story- link the idea of the eco-system to the development of a garden of Eden eco-system city?
Student 2: A new Eden!
Student 3: A smart city but built on the principles of an eco-system?
Student 4: That's a good idea. Nature takes care of all its rubbish better than we do.
Student 3: Yes, nature really has found ways to decompose and recycle everything.
Student 2: Developers should be able to design some better ways of recycling everything we use; we could compost food waste. We could use bacteria on sludge.

Group 2's suggestions for the creation of an eco-city New Eden, based on the principles of an eco-system, provided a timely teaching opportunity-to engage the whole group's attention in identifying the developmental benefits of a development of the forest as a New Eden smart settlement to fulfil local environmental, economic and socio-cultural concerns (Table 4).

Environmental Advocacy: Habitats-In-Boxes

PSTs were tasked to transform their habitat mapping data into 3D habitats in boxes (eco-boxes), to create fictional eco-systems with predator–prey characters representing producers and consumers. In preparation for future teaching placements, they then identified opportunities to teach a range of environmental issues, including the impact of thoughtless consumption, unsustainable development and its contribution to climate change through the creation of play-scripts, newspaper reports and posters.

Table 4 Sustainable development of a smart settlement: New Eden

Pillar of development	Concerns	Mitigations
Environmental	Loss of habitats: trees, land, water sources Loss of bio-diversity and ecosystems, e.g. destruction of newts colony in an area of special scientific interest, loss of nesting area Disruption of animal breeding cycles Litter and waste management-additional landfill sites Increasing urbanisation Increased traffic and deteriorating air quality	Identification and classification of plants, animals and fungi by resident botanist and zoologist Maintenance of exo-systems by forest rangers Management of woodland, coppicing, tree felling and planting. Monitoring of trees Protecting the area of scientific special interest: amphibians, newts Creation of educational trails Maintenance of pathways, cycle tracks, seating
Socio-cultural	Employment opportunities Affordable eco-homes Schools Outdoor learning opportunities Outdoor recreation and leisure, sports and exercise	Recreation, leisure and education Families and group activities Promoting outdoor health activities, exercise and education Forest School outdoor learning for school visits, pupil referral units, Cubs, Scouts, Brownies, Guides Provision of sustainable eco-housing with re-use of materials and sources of energy
Economic	Job creation Business investment Infrastructure Opportunities to lead green technologies developments and renewable energy provision	Employment: forest rangers, botanist, zoologist, education staff, hospitality staff Provision of educational courses Hospitality cafes Construction of New Eden: generating a model of sustainable practice for others to follow, working in partnership with universities and businesses

Representing Those Without a Voice: Predator–Prey Puppets

Using the predator–prey puppets they had created, PST's were positioned as special environmental advocates, preparing to attend the conference in-role as animal and plant delegates. They created conference presentation cards adopting the perspectives of animals and plants at risk because these members of eco-systems are vulnerable to the threats posed by global temperature rises and have no other voice or representation.

Constructing a Board Game as a Teaching Tool

PSTs were commissioned to produce a board game to raise children's awareness of the impact of climate change and the steps that might be taken to reduce or mitigate damage. The game would be presented to delegates at the World Conference and sales of the game would be used to support environmental charities. Working in groups they designed some board games based on a number track, with 3 types of collectable cards corresponding to environmental, economic or social behaviours while identifying damaging or pro-environmental behaviours. Resulting games included: Rainforest Rescue, International Rescue, Earth Watch and Endangered World. PSTs were enthusiastic about creating and using games with primary pupils to consolidate their understanding of sustainable development.

Planning for Placement

The conference, lectures and workshops provided PSTs with an example of holistic and integrated curriculum design (Alexander, 2010) to support their planning for a six-week placement to build their confidence in curriculum-making. Placement expectations were communicated to schools via a placement handbook and a well-attended placement briefing for school-based mentors and host class teachers.

Module Assessments

Post-Placement Group Presentations

In groups of four, each PST presented an evaluation of their resulting planning and schemes of work developed for delivery on placements. PSTs reported that during the module they found the prospect of designing a six-week curriculum quite overwhelming, however once underway and with the support of their school-based mentors, it was an exciting, exhilarating and rewarding way to teach.

Written Assignment

Each PST completed a reflective assignment critically appraising the planning and design of their curriculum-making experiences. Taking responsibility for decision-making and for building curriculum content around the children and PST's interests, was identified as a strength of this approach.

Module Feedback

End of module evaluations were very positive and demonstrated PSTs valued opportunities to make decisions and exercise their judgement about pedagogical choices.

Analysis

How Did the Use of Dramatic Inquiry Raise PSTs Awareness of Environmental Issues?

The contextualisation of science inquiries within dramatic contexts encouraged discussions about pro-environmental behaviours and is an area in which PSTs expressed great interest and would fulfil the GAP for ESD. The discovery of a timid, protected species of newt was unexpected but provided a critical learning opportunity:

> Finding real living newts that only live in this special place was amazing. It did make me think if developers really did decide to bulldoze that part to build houses it might be difficult to re-locate these creatures. Where would you find a similar place locally every space is already developed?

This discovery demonstrated to PSTs that this area was a habitat to a range of creatures-some familiar, some a surprise and, as a habitat the forest provides for the needs of bio-diverse species who are part of a harmonious system.

> Coming across the newts suddenly, just going about their lives, made me realise how vulnerable these creatures are. They've got no ideas how we can take everything away from them.

This reflection raises an issue about human predation of natural habitats as a source of resources for human consumption or use, prompting them to question what right humans have to threaten and destroy? Crist (2014, p. 143, cited in Kopnina, 2014, p. 126) observes:

Human supremacy fuels the top-down conceptualisation of Nature as a resource base, a domain to be used for our ends […] The toxic import of the very idea of resources is unmasked by its normality–a normality instilled by the mode of existence humanity has constructed in accordance with the shared belief in our superiority.

PSTs realised this was an important aspect of sustainable development they could address with children in their future work in schools.

To What Extent Did Participation in Dramatic Inquiry Inform PSTs Understanding of the Three Competing Pillars of Sustainable Development: Environmental, Economic and Social?

Positioning PSTs in expert roles, within a fictitious company, an outdoor setting, and in science workshops provided them with access to more authentic learning contexts. In these contexts, PSTs could, through dialogue, begin to experience and examine the complex nature of sustainable development due to the competing motivations of environmental, economic and social interest groups. Extracts of their discussions indicate their awareness of these competing tensions and the importance of teaching about pro-environmental behaviours:

Student A: I thought businesses would have to anticipate problems in manufacturing and disposal of their products. I'd have thought they'd have to take actions or be fined.

Student B: I think they know where the problems are in production and now they are fined if they breach things like emission quotas or pollute rivers or land. But it's sales that drive companies-once they've sold their products can businesses really be responsible and blamed for people's misuse of them? Look at shopping trolleys- our town used to be full of them until the supermarkets were fined and they chained them all up and made everyone pay to use one. That's not the company's fault, it's people's behaviour.

To What Extent Were PSTs Able to Develop Dramatic Pedagogies of MoE in Their Own Teaching Contexts on School-Based Placements?

During the group presentations that followed student teaching placements, PSTs outline how they had integrated drama or role-play into their curriculum designs. Examples in Key Stage 2 (7–11-year-olds) included the use of role-play to teach the topic of space exploration. One student set up a NASA Mission Control in class in which children took on the role of crew of members of mission control. These roles were conducive to expert framing with everyone working together to prepare for a mission to Mars. Another PST set up a role-play about Jenner's development of vaccination to explore the ethics of carrying out an experiment on a human child to save lives. Groups of children in-role initially re-enacted a scripted version of the historical events but then stepped out of role and adopted positions as scientists, journalists and parents to discuss what should happen if Jenner took the same course of action today.

In Early Years and Key Stage 1 PSTs used traditional tales and fairytales to provide the contexts in which to situate scientific inquiries using role-plays. For example, the story of *The Enormous Turnip* enabled one student to explore forces as pushes and pulls, and *The Big Friendly Giant* (Dahl, 1985) was used to situate a class production line of 'Dream Bottles' to support inquiries about materials and their properties.

PSTs felt being able to link role-play with a story and a subject helped them to integrate different curricular subjects. Those who were willing to immerse themselves in the drama with the children felt more comfortable in attempting to develop this type of approach.

PSTs agreed the use of role-play was an excellent medium through which to explore differing points of view about serious scientific issues with children. However, some students felt constrained by the circumstances they experienced in their placement schools, for example, if there was less freedom to make choices about the design of the curriculum. On a positive note, these PSTs thought that once they gained greater experience in schools, during employment, they would like to try to incorporate drama into their lessons to help children to access and explore difficult concepts in engaging ways.

To What Extent Were PSTs Able to Transfer the Model of Integrated Curriculum Design to Their Own Teaching Contexts on School-Based Placements?

PSTs group presentations, reflective assignments, and module evaluations confirmed that the module had provided valuable professional development to PSTs and they did all attempt to engage in curriculum-making during placement. Within each group presentation, individual PSTs outlined their school contexts, they shared their plans, outcomes and evaluations. A recurring view was their recognition of the value of having a degree of autonomy in decision-making about what and how they chose to teach a topic on placement. In their evaluations, PSTs acknowledged the importance of ensuring the statutory learning objectives and outcomes for each subject were taught effectively through careful monitoring. In science it was felt the use of drama actually enhances the purpose and application of skills, for example, as PSTs worked scientifically to test the quality of water samples in the campus science workshops.

The aim of the module was to provide a model of integrated curriculum design to PSTs for emulation in employment. The module was successful in doing this, however, teacher educators acknowledged the increasing demands placed on class teachers to deliver a crammed primary timetable and how little control PSTs would be granted in taking responsibility for curriculum design whilst on placement.

Conclusion

This chapter has examined how teacher educators deployed the pedagogy of dramatic inquiry to mediate PSTs conceptualisations of sustainable development. Positioning PSTs in fictitious roles as experts facilitated their collaborative knowledge creation through inter-play of their fictional identities and positionalities. In these expert positions PSTs were able to identify, problematise and consider how to mitigate potential damage to the environment. This example of practice accords with the GAP for ESD (UNESCO, 2014b, p.12), as teacher-educators included the key sustainable development issues of bio-diversity and sustainable consumption into their teaching and highlighted the facilitation of collective decision-making to identify future actions.

The use of dramatic inquiry provided a fictitious context in which they could examine real world problems using dialogue and collective decision-making to reach consensus regarding the development of a forest setting. Connecting dramatic inquiry, story-telling, visits and taught subjects enables tutors to address socio-scientific issues (Braund, 2015; Ødegaard, 2003), while providing a model of integrated curricular design (Alexander, 2010). PSTs were supported to recognise their roles as potential agents of change (Ødegaard) responsible for educating the next generation to encourage environmental advocacy and stewardship. In turn, by emulating these approaches, PSTs can now support future pupils to reflect upon the ways they live and the impacts of their choices on the Earth and its diminishing resources (Brundtland, 1987).

Acknowledgements Dr. Debra Kidd, module lead: Gill Peet, content developer; Kathy Schofield, content developer.

References

Alexander, R. (Ed.) (2010). *Children, their world, their education.* Final report and recommendations of the Cambridge Primary Review. Abingdon, UK: Routledge.

Braund, M. (2015). Drama and learning science: An empty space? *British Educational Research Journal, 41*(1), 102–121.

Brundtland, G. (1987). Report of the world commission on environment and development: Our common future. *United Nations General Assembly document A/42/427.*

Budzinsky, F. K. (1995). 'Chemistry on stage'—A strategy for integrating science and school science and mathematics. *School Science and Mathematics, 95*(8), 406–410.

Dahl, R., & Blake, Q. (1985). *The BFG.* Harmondsworth, UK: Viking Penguin.

Dorion, K. R. (2009). Science through drama: A multiple case exploration of the characteristics of drama activities used in secondary science lessons. *International Journal of Science Education, 31*(16), 2247–2270.

Fels, L., & Meyer, K. (1997). On the edge of chaos: Co-evolving worlds of drama and science. *Teaching Education, 9*(1), 75–78.

Heathcote, D., & Bolton, G. (1995). *Drama for learning: Dorothy Heathcote's mantle of the expert approach to education.* Heinemann.

Hodacs, H. (2011). Linnaeans outdoors: The transformative role of studying nature 'on the move' and outside. *The British Journal for the History of Science, 44*(2), 183–209.

Kopnina, H. (2014). Education for sustainable development (ESD): Exploring anthropocentric-ecocentric values in children through vignettes. *Studies in Educational Evaluation, 41*, 124–132.

Louv, R. (2005). *Last child in the woods: Saving our children from nature deficit disorder*. Algonquin Books.

McNaughton, M. J. (2006). Learning from participants' responses in educational drama in the teaching of education for sustainable development. *Research in Drama Education, 11*(1), 19–41.

Mula, I., Tilbury, D., Ryan, A., Mader, M., Dlouha, J., Mader, C., Benayas, J., Dlouhy, J., & Alba, D. (2017). Catalysing change for higher education for sustainable development: A review of professional development initiatives. *International Journal of Sustainability in Higher Education, 18*(5), 798–820. https://doi.org/10.1108/IJSHE-03-2017-0043

Myers, D. (2019). Re-coupling nature and culture: How can primary teacher educators enable pre-service teachers and their pupils to breathe life back into humanity's tin forests? In W. Leal Filho & A. C. McCrea (Eds.), *Sustainability and the humanities*. Cham, Switzerland: Springer.

Ødegaard, M. (2003). Dramatic science. A critical review of drama in science education. *Studies in Science Education, 39*(1), 75–101.

OPAL (n.d.). Citizen science for everyone. *Water survey*. Retrieved from https://www.imperial.ac.uk/opal/survey/watersurvey/

Qablan, A. (2018). Building capacities of educators and trainers. In A. Leicht, J. Heiss, & W. J. Byun (Eds.), *Issues and trends in education for sustainable development*. Paris, France: UNESCO Publishing.

Sjoberg, S. (1998). *Naturfag som allmenndannelse: En kritisk Fagdidaktikk*. Oslo, Norway: Ad Notam Gyldendal.

United Nations Educational, Scientific and Cultural Organisation (UNESCO). (2014a). *Global citizenship education: Preparing learners for the challenges of the twenty-first century*. Paris, France: UNESCO.

United Nations Educational, Scientific and Cultural Organisation (UNESCO). (2014b). *UNESCO roadmap for implementing the global action programme on education for sustainable development*. Paris, France: UNESCO.

United Nations Department of Economic and Social Affairs (DESA). (2010). *Sustainable development*. Paris, France: UNESCO. Retrieved from https://sdgs.un/org

Ward, H., & Anderson, W. (2001). *The tin forest*. Templar Publishing.

Debbie Myers (B.Sc. (Hons), M.A. (Education), M.Sc. STEM Education) is a Chartered Science Teacher (CSci Teach), a senior lecturer, teacher educator and professional development provider at Northumbria University in North-east England. She has previously held posts as a primary Headteacher, Professional Development Lead (Science Learning Centre, Durham University) and as a part-time English Consultant for North Tyneside Local Authority. She designs science education projects and workshops to support teachers' classroom practices. She received a Primary Science Teacher of the Year award in 2007 (AstraZeneca Science Teaching Trust) for creative and innovative practice in science education.

Ice Age is Approaching: Triggering University Students' Interest and Engagement in Gamified Outdoor Playful Learning Activities

Pirkko Siklander and Sari Harmoinen

Abstract Higher education (HE) is required to educate educational experts who can design teaching and learning processes, which trigger students' interest, motivate and engage them. Although research has shown the importance of triggers, particularly in challenging situations, academics still lack practical and theoretical examples on how to design inspiring learning experiences. In addition, little is known about the connection of drama, gamification and engagement in HE learning processes in outdoor contexts. The aim in this chapter is two-fold. First, we describe a gamified role-play design, which was implemented in higher education studies, in an international master's degree program in the forest, and the students studied three concepts—engagement, gamification and collaboration—both by reading scientific articles and through their collaborative experiences in the forest. Second, to analyse students' engagement in the gamified activities, and see which elements positively trigger their interest and engagement in the activities and which on the other hand negatively affect their interest. The data obtained comprises students' self-evaluations throughout the activities and their reflections after the course were considered. The results show that the most important triggers are: (1) collaboration and learning together; (2) role-play activity itself, particularly when it is novel and provides feelings of success; and (3) outdoors activity. Other-directed and intellectual-creative playful triggers were the most evident. Frustration and anxiety can weaken engagement. Implications for educators are discussed.

Keywords Triggers · Engagement · Nature · Playfulness · Drama · Gamification · Role-play

P. Siklander (✉) · S. Harmoinen
Faculty of Education, University of Oulu, Oulu, Finland
e-mail: pirkko.siklander@oulu.fi

S. Harmoinen
e-mail: sari.harmoinen@oulu.fi

P. J White et al. (eds.), *Science and Drama: Contemporary and Creative Approaches to Teaching and Learning*, https://doi.org/10.1007/978-3-030-84401-1_8

Introduction

Designing teaching and learning processes is central to educators' expertise. Designing is based on the theoretical and practical understanding of individual and social learning mechanisms, as well as factors that make learning processes meaningful both for teachers and students. These inspiring and engaging factors can be found from multiple sources, for instance from environments, activities, technologies, other people or contents (Almarghani & Mijatovic, 2017; Renninger & Bachrach, 2015; Siklander et al., 2017). Expert work today in any field includes problem-solving, not only routine but increasingly adaptively. The work requires continuous and flexible knowledge construction and skills for collaboration. Experts use their creativity and playfulness, particularly when solving problems and developing new inventions, methods, technologies, treatments, and procedures (Siklander & Impiö, 2018). In order to develop competences for designing inspiring and engaging learning processes, which have elements of playfulness, games and drama, students need to experience and learn them personally. Therefore, it is important to provide experiences in which students can learn theoretically and experience in practice playful gamified drama (Pajari & Harmoinen, 2019). We use the term gamification to describe the way learning is made playful and exploratory for adult learners, through making it like a game. We understand that drama is much more than games, but that games and drama are connected, sharing many common elements (Winston & Tandy, 2008). When made like a game, adult learners as role-players can experience an imagined world as if it is real, but without real-life consequences. Games and gamified learning have been studied among adults (e.g. Korkealehto & Siklander, 2018), but gamification by using outdoor environments in higher education has not gained scientific attention.

The aim of this chapter is two-fold. First, we describe a gamified design, which was implemented in higher education studies, as part of an international master's degree program. The design was implemented in the nature, and the students studied three concepts both by reading scientific articles and through their participation in collaborative experiences in the forest. The concepts were: engagement, gamification, and collaboration. Secondly, to analyse students' engagement in the gamified role-play activities, to see which elements positively trigger their interest and engagement in the activities and those that have negative impacts on these indicators. Implications for educators are discussed.

Gamification, Playfulness, and Drama

Most of the studies have been conducted in digital settings, indoors, and the physical activity involves sitting at a desk (e.g. Korkealehto & Siklander, 2018; Sun et al., 2018). Although there are differences among gamified applications, gamified course design and related applications, gamification can enhance student engagement, foster language learning, and offer positive learning experiences. Gamification itself is not an adequate element for triggering students' interest and engagement, but strategies to support collaboration and to create a positive atmosphere are needed to cultivate the gamified learning process (Korkealehto & Siklander, 2018). In addition, scaffolding, collaboration, and perceived ease of use seemed to be triggering elements in gamified learning environments (Sun et al., 2018). We consider gamified learning processes as a form of active learning, which requires students' participation in-role not as consumers, but as active and proactive producers. When gamified processes are designed well and used correctly, they have potential to improve learning (Aldemir et al., 2018). Having this in mind, we focused on developing opportunities for effective collaborative learning, problem-solving and engagement. In addition, to facilitate students' gaming competences we combined a set of knowledge, skills and behaviour that can support students to perform certain tasks to a given level (Che Pee, 2011).

When designing gamified and playful drama, we also considered game characteristics versus 21st century skills defined by Romero et al. (2015), and characteristics of collaborative gamification (Parzhetskaya, 2014). Many studies (Jackson, 2002; McNaughton, 2006) have shown the potential of drama to provide a more concrete frame of reference in situations where participants have different challenges to participate in learning. Accordingly, *competition* and *goal-setting* are the first characteristics. Competition is enjoyable and it triggers motivation for learning and completing the task. *Complex collaboration* affords students' possibilities for developing flexible knowledge structures together. Collaboration takes place within knowledge, information analysis and leadership as well as joint efforts. Collaborative problems should be increasingly difficult and present continued challenges. Continuous challenges require new skills and new learning from the team, however, teams have to monitor, control and reflect on their actions. Mistakes and conflicts during the game should encourage students to elaborate new strategies and new ways of collaboration. Also, success is shared; success means achievement for all students.

Secondly, games and play always follow *rules*, and the students negotiate and agree about them. Jointly agreed rules and communication systems can advance engagement and progression. Another characteristic encourages *choice*: choosing an avatar, designing an identity, and choosing strategies and understanding errors. *Fantasy* is one characteristic which denotes fidelity, context, and challenge. The context includes the settings, narrative story, imagination, and scenarios.

The content of the gamified learning process is important. Many tricky and difficult themes, as well as multidimensional content can be incorporated in gamified learning processes. STEM (Science, Technology, Engineering and Mathematics) or STEAM (Science, Technology, Engineering, Arts and Mathematics) themes create different possibilities and so can provide excellent gamification circumstances (Özsoy & Özsoy, 2018). Traditionally STEM means integration within science, mathematics, engineering and/or technology, which has led to approaches such as STEM education in the United States (Czerniak & Johnson, 2014). Recent changes have resulted in the integration of art with STEM facilitating STEAM education (Sousa & Pilecki, 2013). Integrating art includes different forms of art in multiple ways such as literature, drawing/painting, drama, music, and sculpture (Reif & Grant, 2010).

In the present case, learning contents included were engagement, gamification and collaboration, but also the forest ecosystem, or more precisely nature environments as an affordance system, where students perceived and understood nature for different purposes. To support understanding students used technologies to search and locate information in the nature. Nature provides a variety of physical phenomena that prompts wonder and curiosity, particularly for students, who lack such experiences.

As gamification may be viewed as a pedagogical design of *playful activities*, it provides students with opportunities to use their creativity. It is important to define play, playfulness, and drama. *Play* is an activity and playfulness denotes the quality of the activities (Hyvönen, 2008; Resnick, 2018). Teachers use multiple forms of play in elementary school contexts (Hyvönen, 2011) but what kinds of play or playfulness are students in Higher Education (HE) ready to engage in? These students come from different countries and cultures, and the idea of playfulness in HE may be odd for them. Therefore, it is crucial that teachers who implement playful learning processes, are engaged and confident in using it, because otherwise students will lack interest and joy (Kangas et al., 2017). Playfulness covers embodying activities, meaning hands-on, mind-on and body-on activities requiring students to draw upon higher order thinking skills.

Drama can be seen as a game, and game as drama, if we give freedom for students to use their imagination and knowledge, to create and improvise, and to be open to their opinions and perceptions by playing and acting. Drama education has been studied in the field of science and found to be an effective learning strategy for learning sciences (Henry, 2000; Ødegaard, 2003). Narrative text is more common in the life experience of learners than expository text in organising knowledge, use of drama can lead to better empathy with science and more effective cognitive learning (Braund, 2015). Students in different educational levels, also in HE, need to develop their thinking skills through investigation, exploration and questioning (Alexander & Flutter, 2009; OECD, 2017; Ofsted, 2008). Through drama and gaming students can practice these skills. Drama integrates fact and fiction and can promote empathy and a caring attitude when dealing with the issue of global injustice (Lehtonen, 2015). By applying drama we can evaluate what values and facts influence our actions. Gamified activities and drama afford situations where students can practice decision-making and negotiation about values and practices at personal, interpersonal and global levels (Ødegaard, 2001, 2003).

Triggering Interest and Enhancing Engagement

Interest, motivation and engagement form a process, in which triggers have a key role, because they can awaken and maintain student's interest (Hidi & Renninger, 2006; Järvelä & Renninger, 2014; Renninger & Bachrach, 2015). Triggers are any elements or moments, which students find catching, maintaining and returning their interest. Triggers provoke feelings, such as excitement, joy and curiosity, which makes students willing to continue activities. Triggers include affective and cognitive components, and they are motivational structures. The problem, however, is to know which triggers are meaningful for students and effective enough to keep students' attention and activities dynamic in order to achieve well developed individual interest (Hidi & Renninger, 2006).

The four-phase model (Hidi & Renninger, 2006) of interest development suggests that each phase can be characterised by varying amounts of affect, knowledge and value. The four phases are: (1) situational interest, (2) maintained situational interest, (3) emerging individual interest, and (4) well developed individual interest. Teachers and educators often think that students either do or do not have interest, but they might not recognise that interest can be aroused by features of environments, by designing teaching and learning and by different activities. Arousing interest takes time, practice and awareness for perceiving triggers around students' formal and informal, virtual and physical environments.

Only a few studies have explored triggers in higher education. A comparative study in an online course shows that reflective triggers were extensively employed by the participants and they were perceived to be useful for reflection and learning (Veerporten et al., 2012). Our earlier findings (Siklander et al., 2017) suggest that the *topic* is the most effective trigger. It raises interest first at the individual level and maintains a triggering effect at the collaborative level. A triggering topic is open and requires problem-solving. Another effective trigger deals with *collaboration and interaction*, which refers to reciprocal and equal participation. The third trigger is *feedback*. Peer feedback motivates more than teachers' feedback. Other salient triggers are emotional atmosphere, collaborative learning, learning environment, teacher and goal-orientation. Because triggers can advance problem-solving and increase enjoyment for learning. (Roberts & Ousey, 2004), we introduce playful triggers, which we consider as positive triggers. However, triggers can be also negatively experienced, when they hinder motivation and engagement, as well as problem-solving and enjoyment.

Playful Triggers

Proyer et al. (2018) have defined five categories of adult playfulness, which we (Siklander & Kangas, 2020) have adapted, and divided into five playful triggers. On the basis of the aforementioned work, we adapt playful triggers into the collaborative learning context of this study with students in higher education. The triggers are slightly overlapping, and one trigger can boost another.

Global playfulness recognises playfulness as a global phenomenon, and international students as individuals and representatives of their cultures, enacting playful activities within the frames of their cultures and everyday situations. International students orient between their original cultures, their peer's cultures as well as Finnish cultures in their daily studying activities. While doing so, global playfulness invites students to experience diverse playful situations as positive, interesting and mentally inspiring. Global playfulness denotes the fact that there are no right or predefined ways to realise playfulness, but it can exhibit many different forms. Students and teachers could therefore discuss the features of global playfulness occurring in outdoor drama settings, how they can trigger engagement and how to increase their understanding of it, and what they find particularly inspiring about global playfulness when students are experiencing gamified activities in the natural forest environment.

Playfulness which is other-directed suggests that participants recognise and utilise the potential of playfulness that emerges in their interaction: interaction between students, and interaction between students and environments. Playful students contribute to a free, relaxed atmosphere and support and encourage other students. They enjoy fooling around and playing together, but at the same time they focus on joint goals and shared activities. Collaboration itself, when it is successful, is an effective trigger to encourage collaboration (Siklander et al., 2017). Drama and role-play are natural ways for developing other-directed playfulness. In groups, students can perceive and reflect how one's playfulness and idea generation affects other students. This kind of mirroring role can be even one of the selected roles in drama. Because playfulness requires contents and contexts, the subject matters as plotted narratives are useful. Other-directed playfulness can be practiced in different team activities, for instance in work life.

Light-hearted playfulness denotes students' abilities to relate joyfully, confidently and without anxieties to present and new approaching situations. They can improvise and act spontaneously without wasting energy on worrying about the consequences of their behaviour. In the higher education context, light-hearted playfulness may relate to students' ability to live in the moment and use conscious playfulness in different situations. For instance, in problem-solving, playfulness can release cognitive resources and in drama it is considered to be an expected trigger. Light-hearted playfulness is also connected to health and well-being, showing that students can be open and curious to new experiences and creativity (Antinori et al., 2017).

Intellectual-creative playfulness refers to creative (divergent) thinking in challenging problem-solving situations, in which students enjoy playing with ideas. Within a group each student tries to act as an example of how to perform intellectual-creative playfulness. For instance, finding novel solutions in changing situations and contexts. Intellectual-creative playfulness can be developed by throwing oneself into different authentic problem-solving situations and approaching them as unique learning possibilities.

Whimsical playfulness means being open to novel ideas and deviation from the usual, habitual and ordinary ways of responding. As an example, whimsically playful students do not want to explore study contexts as they usually do, instead, they are ready to break the rules, changing their behaviour to develop something new. They can invent seemingly odd or juxtaposed ideas and they try to reach uncommon solutions. These students are interested in other whimsically playful individuals and can support them.

Engagement

Engagement is a complex, multifaceted construct including overlapping cognitive, emotional and behavioural dimensions (Boekaerts, 2016; Eccles, 2016; Fredricks et al., 2004). This classification is well-accepted, but not well-theorised therefore, dimensions are unclear and challenging to measure (Eccles, 2016). Also, the fourth dimension, agentic engagement, was suggested, which refers to proactive behaviour.

Engagement is seen as an individual's response to contextual features (Fredricks et al., 2004). In our context we consider engagement from a collaborative problem-based approach: how do students respond to obstacles, challenges and problems? Do they take them as learning opportunities and invest effort to solve them, or do they self-handicap and quit? *Collaborative engagement* (Järvelä et al., 2016) refers to interaction, particularly types of interaction in self-regulated learning design. We claim that while engagement is socially constructed, and therefore dynamic and malleable (Lawson & Lawson, 2013), in collaborative engagement the entire group has power to effectively impact on engagement.

There is a debate among researchers whether engagement is a state-like construct or whether it has more stable characteristics (Eccles, 2016). We suggest that engagement should be seen as a process. The process is not straight forward. Therefore, in addition to short-term and moment-to-moment exploration, research should also focus on longitudinal processes and studies of different age and cultural groups, and in different learning environments and learning models, such as gamified designs. One challenge in the engagement process is to define accurate indicators for illuminating student engagement in their tasks and learning. Indicators can be visible or invisible or they can be accessible or not-accessible. For instance, involvement in activities at hand shows behavioural engagement (Payne, 2017), but the inhibition of cognitive and emotional engagement is not so obvious. Observations and self-reports are mainly used to capture the level and type of engagement. Self-reporting also raises the question of whether students understand what we mean by engagement, therefore we suggest that it is beneficial for them to know about the phenomenon on which they are reflecting. We know that teachers' engagement and enthusiasm (Kangas et al., 2017) and the emotionally safe atmosphere they provide, without power relations (Eteläpelto & Lahti, 2008), accompanied by collaboration and scaffolding (Korkealehto & Siklander, 2018; Sun et al., 2018), seem to be grounding elements for triggering students' interest and enhancing engagement.

Nature as a Learning Environment

Current research evidence suggests that younger and older students experience activities in natural environments positively (Kangas et al., 2018). In educational sciences, there is a growing need for further research on how outdoor environments and nature can be used in education (Gilbertson et al., 2021). Some studies have shown (e.g., Skinner & Chi, 2012; Alon & Tal, 2015; Lekies et al., 2015), students report high levels of student engagement and motivation, during both student-elected and school-mandated nature activities.

Nature as a learning environment has been studied widely in early childhood contexts (Siklander et al., 2020), but seldom with adults. However, environmental characteristics that are important for children, can be similarly important for higher education students. For instance, diversity, variation, formlessness and manipulation are inherent only in natural landscapes (Fjortoft & Sageie, 2000; Nabhan & Trimble, 1994). Kirkby (1989) concludes that natural settings have a degree of complexity, plasticity and manipulability. Affordances highlight utility and flexibility of the natural environment, with nature providing possibilities for students to experience playful gamified drama activities and cultivate playful triggers in their interaction and also collaborative learning.

The theory of affordances (Gibson, 1979; Hyvönen, 2008) brings to light how to perceive nature as a learning environment in higher education. Affordances, prominently discussed in research regarding the relationships between students and nature, is the range of actions that the environment or environmental objects can provide to students (Fjortoft, 2004). The concept of affordances is reciprocal and conditional, it includes nature, the student and their interaction; meaning that the affordances are unique for each individual, depending on their goals, needs, ability to perceive, and ability to use the perceived affordance (e.g. while one student saw a dog in the nature as a funny companion and used the dog as part of the game scenario; another saw it as a wolf to be frightened of and avoided). That means, affordances are not in the environment nor in the individual but can be realised in their interactions. Students find affordances which they are capable to find and realise, and which are meaningful for them.

The theory of affordances regarding playful and gamified learning (Hyvönen, 2008) identifies additional attributes to affordances. Succeeded affordances are those which students can realise; they are based on their previous knowledge and experiences ('I know this already'). Another attribute, expected affordances, are those which students find meaningful when playing or learning ('I want and I need this'). Hidden affordances are those which are completely unknown to them; but once they invent them in their exploration, such as in the complex affordance networks nature provides, they have learned something new that can be drawn upon going forward.

In our case-study the main learning goals were: (1) to understand engagement by combining its theoretical and practical aspects; (2) to learn how to thrive in nature; and (3) to apply theoretical knowledge about gamification into practice. For this third goal students collected nature items, took photographs, and created podcasts and videos from their adventurous experiences in nature for resources to later use when designing a game scenario for children. The activities in nature required students to be constantly active in perceiving affordances to be used in the game scenario, as well as students' interaction and collaborative knowledge construction. The 'nature' comprised of a Finnish summer cottage at a lake and the surrounding forest of large pines, spruces and birch trees. Mostly the forest was without any pathways and was somewhat difficult to walk through because of wet, swampy areas and thick vegetation. Nature was not at its best because of the late-September weather, which was rainy, cold and windy. At the cottage we had shelters and a fireplace.

Description of the Case

The aim of the study was to explore students' behavioural, cognitive and emotional engagement in the activities, as they self-experience and evaluate. Participants stage their imaginary images of themselves and others in Ice Age. The participants were international students (N = 12) in the educational master's degree program. The case was a part of the international HE course 'Problem-solving case 2' (10 credits). Three teachers participate in the Ice Age day, their role being to observe activities, provide essential facilities (food, materials and instructions), and also assist students if needed.

Pedagogical Design

The process of the Ice Age gamified drama followed pedagogical design, which is based on the playful learning process model (Hyvönen, 2008; Kangas, 2010), including orientation, play/game, and an elaboration phase. Theoretically the model draws on the learning sciences, play research and features of playfulness (Hyvönen, 2008; Kangas, 2010). It follows drama and role-play approaches: warm-up, action and sharing (Rassmussen & Khalich, 2000) (Fig. 1).

The purpose of the *orientation* is to introduce students to the topic and activate their prior knowledge and experiences. Further, it is important to activate students' interest and their readiness for collaboration. Orientation also includes understanding the task, setting goals and planning strategies. Orientation took place indoors at the university, and it included orientation to the process of problem-solving and to the imaginary narrative of the Ice Age adventure. It was important to highlight the meaning of the collaborative problem-solving and differences between a task and problem-solving. The latter requires a rich environment and accepting frustration and failure as a part of the process. Solving open problems allows agency, autonomy, creativity and ownership and is different to just completing well-structured task. A brief lecture about theoretical constructs, such as engagement, collaborative engagement, gamification, game characteristics, and collaborative gamification were discussed with the students.

The Ice Age case was introduced with an imaginary context, problem, plan and adventure. The students were randomly divided into groups and the groups in this setting were called 'clans'. Each clan was named after wild berries which can be found in the forest : bearberries, crowberries and lingonberries. In addition to describing

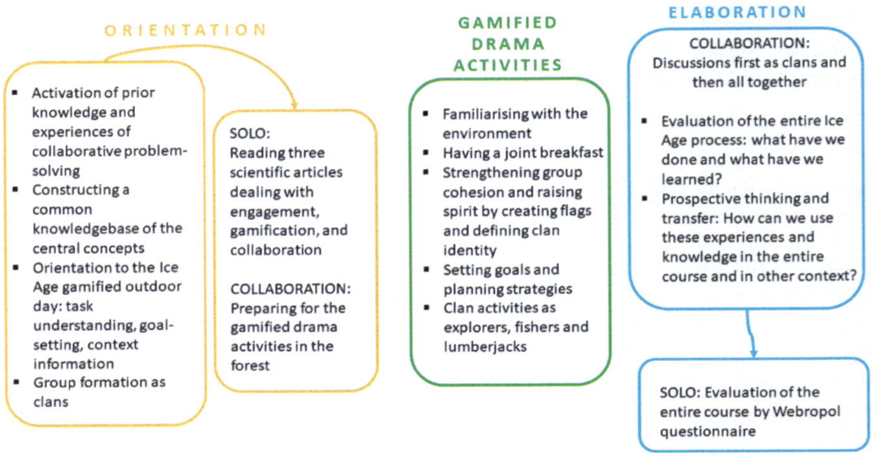

Fig. 1 Pedagogical design of the ice age

the forthcoming outdoor day, students were informed about other important things to take into account. For instance, safety and hygiene issues, appropriate clothing, and useful equipment were discussed.

Gamified Drama Activities

The imaginary context is described as a narrative 'Ice Age is approaching', and the clans have to prepare themselves for it. Students engaged in role-play and were required to take part in three different activities: (1) find a new location for the clan, (2) provide food for the clan, and (3) ensure energy resources for the clan. The narrative implies a problem, but in addition, the actualisation of each role was complicated, because the environment was totally unknown, and the complexity was increased by learning goals: understanding engagement and gamification, and to create a game scenario on the basis of their understanding and activities. The problem was introduced as a sequence of the following three narratives:

> Each story began by... "The ice-age is predicted to become true, and you as a clan (Lingonberry, Crowberry and Bearberry) are responsible for your own survival. There are enormous threats to acknowledge and to prepare for."

(1) EXPLORERS: You cannot stay at the Noiteli's cottages, but you are forced to look for other locations to live. Your clan adopts the role of explorer and starts a long and exhausting journey to better living areas. Before leaving, equip yourselves for the unknown journey. You finally find a lean-to (Laplander's hut) and decide to base your clan there. Look for the environment and find affordances: what it provides, either good or bad, for your clan.

(2) FISHERS: Your clan has realised that the stored food, such as fish, berries, roots, game and herbs, is not enough for approaching difficult years. You have no other choice but to equip yourselves, make plans and go fishing. Where, what and how? Finally, you are rowing on the huge lake with changing conditions.
(3) LUMBERJACKS: Your clan realises that the ice-age is a long, long winter, and enormous amounts of wood will be needed to ensure warmth. In addition, chopped wood is needed for cooking and heating water for washing humans and clothes. There is no other option than to organise a 'labour camp'. Not only the strongest members, but everyone, has to do their bit for the community.

In addition, clans were asked to collect resources and plan a new game scenario, and students were asked to complete a pre- and post-evaluation form. In the form, they estimated the level of behavioural, emotional and cognitive engagement by drawing a line on the scale up to 100.

Students arrived at the Noiteli cottage in the morning, and one student said: "Already the drive here was an adventure". The drive was one hour including 4 kms of muddy road in the forest. For most students the forest environment was totally new. We had our first coffee, bread, fruits and soup, and settled around. Each clan created a flag and started to make their plans and each took on one of the three roles to execute. Explorers orienteered through the forest to the Laplander's hut and found an open fireplace and sausages there. They lit the fire, fried sausages and as in each scenario, they designed how to use this experience for the new game scenario. They created podcasts, took photos and videos, and collected natural items. Because of stormy weather, fishers could not go to the lake, they enacted the activities differently. They familiarised themselves with the fishing equipment and fishing practicalities. They looked for additional information on the internet. They wore safety jackets and boarded the boat with the fishing equipment. Lumberjacks found an axe and pile of logs, which they started to chop into smaller pieces of firewood. They organised their work: one gave a log to the chopper, one chopped it, and one stacked the pieces. Meanwhile the students had food and drink available. The day ended with the joint elaboration, in which everyone reflected on the entire day. In this phase, the students were rather exhausted after a long and intensive day outdoors.

Results: Students' Interest and Engagement

We were interested in students' perspectives on gamified outdoor learning, particularly those elements which can trigger students' interest and enhance their engagement. We looked for positive and negative triggering elements and moments. Students' interest and engagement were analysed at two different points in the course, first during the outdoor activity day (55 responses), and second, at the end of the entire course when students responded to an online-questionnaire with Likert evaluations and open questions. As a result of this case the most obvious element to trigger students' interest and engagement were *collaboration*, *activities* and *outdoor gamification,* which we discuss next.

Collaboration and Challenging Situations

Collaboration seemed to be the most effective element for triggering students' active participation together in the gamified outdoor activities. During the outdoor day, students reported on the importance of their teammates, on how they held fruitful discussions, provided encouragement for each other, and constructed knowledge together. Collaboration can be analysed by using three dimensions: (1) feelings of togetherness, (2) cooperative, and (3) collaborative knowledge construction (Dukuzumuremyi & Siklander, 2018). These were integrated in students' comments, for instance the following excerpt illustrates togetherness and collaborative knowledge construction: "We kept encouraging each other and also brainstormed together." The meaning of togetherness was experienced by one clan when one member could not participate in person but attended using technologies. "Also a student who was located overseas encouraged us from Spain and made us feel good." And, "another teammate participated online and actively also created a sense of belonging and group relatedness".

Shared success, which is typical in gamification and desired in successful collaboration, was celebrated by the clans: "Happy we finally manage everything! Our teammates are supportive, which makes me feel optimistic." However, active collaboration is not always easy, and it requires negotiations: "It is getting more difficult! And we decided to re-design the game, because we realised that we must combine three activities together. I feel less excited, and we all have different ideas." Another excerpt recounts the meaning of collaboration when facing challenges: "We get lost and waste some time. It makes me a little nervous 'cause I am worried if we could complete the goals in time, but the teammates are nice and we overcome [sic] those difficulties." Problems are not negatively affecting elements, as one student concluded: "Despite all problems, this was the most collaborative group work I ever experienced."

Several students highlighted the ways they collaborated in challenging collaborative learning situations: "In exploring the location, we had to discuss and agree a lot about where the suitable new 'home' was. There was a lot of interaction among the group; navigating through the forest the first [time] without any compass, was challenging." Knowledge construction is based on previous knowledge and experiences, which the students applied to their activities: "I have activated my prior knowledge regarding fishing. Even though I have not fished in Finland, I have already done fishing [sic] back in Africa. Those experiences have been useful in helping our group to come out [sic] with fishing for survival." Despite the many challenges, the collaboration in clans was fruitful: "Had some difficulties collecting info and had a nice but intensive discussion, I enjoyed it. Let's move!". The findings of the questionnaire are in line with the above findings, however, it reveals that after the gamified outdoor day, the clans continued their collaboration in increasingly efficient and professional ways.

Activities and Success with Novel Experiences

The activities outlined above triggered students' interest and engagement. Chopping wood together was surprisingly inspiring, and although only one member was chopping, they could make it a collaborative activity by sharing the task and turn taking: "Lumberjack sounds simple, but the game has to be educational, fun and challenging. We have to take collaboration into consideration as well." Successful experiences inspired the students, and one student wanted to do it again and again. She said "This is addictive to me." Other students expressed their positive feelings: "Happy to learn how to chop logs", and "Excited that I made it." "Chopping wood was a great experience because we are able to actually do it. I felt accomplished and satisfied by doing it." The Lumberjack's role also inspired imagination: "This activity reminds me of some Netflix scenes:)". Another student raised the meaning of imagination in the activity-based learning situation: "I enjoyed the activity, even if we didn't go fishing. It was nice to imagine and plan the conditions:) I love the activities:)" Chopping wood was a totally new experience for the students: "Excited about the activity and I have not tried this before." Other activities, which were novel to the students and in which they succeeded, raised their interest, for instance, "Successful in lighting the fire".

Outdoor Gamification and Learning with Drama

The present case includes gamified elements, such as imaginary context, problems, planning and the adventure. Drama could be utilised to promote a more informed understanding of the way scientific knowledge is generated (Ødegaard, 2003). Although we did not include competition elements, the focus was on collaboration, one student pondered the effects of competition on engagement: "If two teams are doing the same activity, somehow you will compare with each other. If you lose, it will affect your engagement for a while." On the basis of their preferences, students pondered what kinds of games could enhance learning and how. They preferred outdoor games, because these required several senses to discover learning, activating and stimulating deep thinking and reflection. This was evidenced in practice, when students were chopping logs: "I think we can teach something related to physics with this type of activity. 'How to convert wood into energy?'" Students preferred games that trigger non-intended learning situations (e.g. problem-solving), provide meaning, empowerment, accomplishment, feelings of ownership, and social influence, and that are unpredictable. They favour games that integrate theoretical content with roles, which can enable students to be imaginative and creative. As the students concluded, triggering games provoke students' thoughts, this happened during the outdoor gamified day. It raised students' interest in possible historical contexts: "I am thinking about the prehistory era, of the first people in the world and how they made tools from wood, stones and so on".

Students' Evaluation of Their Engagement

Overall, during the gamified outdoor activities students evaluated their engagement level rather high. The averages of their estimated emotional engagement was 74%, behavioural 73% and cognitive 72%. In addition to evaluating levels of engagement, they wrote comments to justify their evaluations. Negative elements were explained by tiredness and fatigue, hunger, fears, bad weather and time limitations, as they expressed: "Time is limited. I am a bit anxious. I want to do and design something fun and great", and "At the end of the first task we were somehow cognitively disengaged because of the tiredness, fatigue and hunger". The online questionnaire did not bring additional information about negative effects, however, some expectations of more constructive and respectful interaction was demanded by one student.

At the end of the problem-solving course the students evaluated their past engagement and reflected on its usefulness for their future career: "Personally, I am motivated to employ all the new skills that I have learned and acquired in my future career." The level of engagement was discussed in relation to each clan's level of satisfaction, and to examine how students purposefully steered their behaviour to enhance clan members' engagement. Although students describe mostly behavioural engagement, emotional and cognitive dimensions are also analysed, for instance frustration and anxiety can weaken emotional engagement, and cognitive engagement enables students to search for information. Responses show that behavioural engagement is the easiest to analyse, regulate and predict, as one student wrote: "I knew from the beginning how my behaviour was going to be, but I could not know how my cognition or emotion would be". Engagement as a process was not stable, instead it wavered including downturns and building up phases. However, metacognitive abilities, understanding and awareness of other students' understanding deepened during the entire course, and the gamified outdoor activities triggered it and afforded a good basis for it.

Conclusion and Discussion

This study provides an example of how to design a gamified and playful learning process, which integrates theoretical concepts and collaboration in outdoor activities. The drama can be seen in this case as a way of acting and creating a feeling of real life. The activities took place at the summer cottage, in the forest and at the lake shore. The plot of the narrative was triggered from the angle of climate change and integrated historical elements. This provided an excellent way into understanding climate change as a phenomenon and increase participant understanding in the way students are looking for (Harmoinen et al., 2020). The students explored as clans, and their task was to prepare their clans for approaching the Ice Age. The results show that the most important triggers are collaboration and learning together, the activity itself, particularly when it is novel and provides feelings of success, and the gamification outdoors.

The students had studied collaboration in their master's program, and this case evidences their capability to understand it at a conceptual level, and to apply it in their activities in a totally different context. Their emphasis on collaboration was also seen in the questionnaire responses, where questions were focused on individuals, and the students answered from the clan's perspective, using 'we' more than 'I'. Collaboration in the forest activities turned out to be challenging, and perhaps therefore engaging, when activities were experienced successfully. This is in line with our earlier results: collaboration is a salient trigger for more collaboration (Siklander et al., 2017). This notion calls for more focused empirical studies about collaborative engagement in diverse higher education gamified contexts: how do we engage the entire group: behaviourally, emotionally and also cognitively?

Since the process heavily emphasised theoretical content, the role of playful triggers (Siklander & Kangas, 2020) is worth considering. Playfulness triggered students to integrate their own cultures and Finnish summer cottage culture with academic learning cultures and imaginative narrative and context. Integration was seen, for instance, in the fishing episode. Other-directedness was the most evident playful trigger, since collaboration took place in many ways, it relaxed the atmosphere, and supported and encouraged other students. Light-hearted playful triggers were not seen in the data. This may be due to the students' aim to do their very best to complete the day and to accomplish the goals, and also due to negative effects, such as tiredness and time restrictions. Intellectual-creative playful triggers, instead, were plentiful, those relating to problem-solving and other challenges in nature, and the task of creating a game scenario based on their own experiences. Talking about science became natural, it took courage to present one's own understanding, their own prior information was examined reflectively. The flags, which the students created at the beginning, is proof of this kind of playful trigger. Whimsical playful students were willing to explore totally new study contexts. They were ready to make new rules, change their behaviour and develop together a new learning culture.

Nature as a learning environment affords experiences which involve whole body and whole clan participation. Nature occupies students behaviourally, emotionally and cognitively, and how they take advantage of it depends on their levels of ability and engagement. Students who found a dead snake could utilise it as a resource in game scenarios as well as for making jokes and relaxing the atmosphere.

Acknowledgements The course and the adventure day was developed in collaboration by the teachers' team. We acknowledge each and everyone's spirited contribution to the success of this unique education effort. Thank you to Niina Impiö, Essi Vuopala and Tiina Salmijärvi.

References

Aldemir, T., Celik, B., & Kaplan, G. (2018). A qualitative investigation of student perceptions of game elements in a gamified course. *Computers in Human Behavior, 78*, 235–254.

Alexander, R. J., & Flutter, J. (2009). *Towards a new primary curriculum: A report from the Cambridge primary review. Part 1: Past and present*, Cambridge, UK: University of Cambridge Faculty of Education.

Alon, N. L., & Tal, T. (2015). Student self-reported learning outcomes of field trips: The pedagogical impact. *International Journal of Science Education, 37*(8), 1279–1298. https://doi.org/10.1080/09500693.2015.1034797

Almarghani, E. M., & Mijatovic, I. (2017). Factors affecting student engagement in HEIs-it is all about good teaching. *Teaching in Higher Education, 22*(8), 940–956.

Antinori, A., Carter, O. L., & Smilie, L, D. (2017). Seeing it both ways: Openness to experience and binocular rivalry suppression. *Journal of Research in Personality*, 68, 15–22. https://doi.org/10.1016/j.jrp.2017.03.005

Boekaerts, M. (2016). Engagement as an inherent aspect of the learning process. *Learning and Instruction, 43*, 76–83. https://doi.org/10.1016/j.learninstruc.2016.02.001

Braund, M. (2015). Drama and learning science: An empty space? *British Educational Research Journal, 41*(1), 102–121. https://doi.org/10.1002/berj.3130

Che Pee, N. (2011). *Computer games use in an educational system* (Doctoral dissertation, University of Nottingham).

Czerniak, C. M., & Johnson, C. C. (2014). Interdisciplinary science teaching. In N. G. Lederman & S. K. Abell (Eds.), *Handbook of research on science education* (Vol. 2, pp. 537–559). Routledge.

Dukuzumuremyi, S., & Siklander, P. (2018). Interactions between pupils and their teacher in collaborative and technology-enhanced learning settings in the inclusive classroom. *Teaching and Teacher Education, 76*, 165–174.

Eccles, J. S. (2016). Engagement: Where to next? *Learning and Instruction, 43*, 71–75. https://doi.org/10.1016/j.learninstruc.2016.02.003

Ernst, J., McAllister, K. Siklander, P, & Storli, R. (2021). Contributions to sustainability through young children's nature play: A systematic review. *Sustainability, 13*, 7433. https://doi.org/10.3390/su13137443

Eteläpelto, A., & Lahti, J. (2008). The resources and obstacles of creative collaboration in a long-term learning community. *Thinking Skills and Creativity, 3*(3), 226–240. https://doi.org/10.1016/j.tsc.2008.09.003

Fjortoft, I. (2004). Landscapes as playscape: The effects of natural environments on children's play and motor development. *Children, Youth and Environments, 14*(2), 21–44.

Fjortoft, I., & Sageie, J. (2000). The natural environment as a playground for children. *Landscape and Urban Planning, 48*(1/2), 83–97.

Fredricks, J. A., Blumenfeld, P. C., & Paris, A. H. (2004). School engagement: Potential of the concept, state of the evidence. *Review of Educational Research, 74*(1), 59–109. https://doi.org/10.3102/00346543074001059

Gibson, J. J. (1979). *The ecological approach to visual perception*. Houghton, Mifflin and Company.

Gilbertson, K., Bates, T., Siklander, P., & Ewert, A. (2021). *Outdoor Education* (2nd ed.). Human Kinetics.

Harmoinen, S., Koivu, K., & Pääsky, L. (2020). University students' readiness for social activity in climate actions. *Discourse and Communication for Sustainable Education, 11*(1), 134–152. https://doi.org/10.2478/dcse-2020-0012

Henry, M. (2000). Drama's ways of learning. *Research in Drama Education, 5*(1), 46–62.

Hidi, S., & Renninger, K. A. (2006). The four phase model of interest development. *Educational Psychologist, 41*(2), 111–127.

Hyvönen, P. (2008). *Affordances of playful learning environment for tutoring playing and learning (Doctoral dissertation)*. University of Lapland Printing Centre.

Hyvönen, P. (2011). Play in the school context? The perspectives of Finnish teachers. *Australian Journal of Teacher Education (AJTE), 36*(8), 5.

Järvelä, S., Järvenoja, H., Malmberg, J., Isohätälä, J., & Sobocinski, M. (2016). How do types of interaction and phases of self-regulated learning set a stage for collaborative engagement? *Learning and Instruction, 43*, 39–51. https://doi.org/10.1016/j.learninstruc.2015.10.006

Järvelä, S., & Renninger, K. A. (2014). Designing for learning: Interest, motivation, and engagement. In D. Keith Sawyer (Ed.), *Cambridge handbook of the learning sciences* (2nd ed., pp. 668–685). New York: Cambridge University Press.

Jackson, T. (Ed.) (2002). *Learning through theatre: New perspectives on theatre in education* (London, Routledge)

Kangas, M. (2010). *The school of the future: Theoretical and pedagogical approaches for creative and playful learning environments (Doctoral dissertation).* University of Lapland Printing Centre.

Kangas, M., Siklander, P., Randolph, J., & Ruokamo, H. (2017). Teachers' engagement and students' satisfaction with a playful learning environment. *Teaching and Teacher Education, 63*, 274–284. https://doi.org/10.1016/j.tate.2016.12.018

Kangas, M., Vuojärvi, H., & Siklander, P. (2018). Hiking in the wilderness: Interplay between teachers' and students' agencies in outdoor learning. *Education in the North (Special Issue Perspectives on Spaces for Teaching and Learning), 25*(3), 7–31. Retrieved from https://www.abdn.ac.uk/eitn/documents/Vol_25_issue_3/EITN_2018_03_02_Kangas.pdf

Kirkby, M. (1989). Nature as refuge in children's environments. *Children's Environments Quarterly, 6*(1), 7–12.

Korkealehto, K., & Siklander, P. (2018). Enhancing engagement, enjoyment and learning experiences by gamification on an English course for healthcare students. *SeminarNet, International Journal of Media, Technology, and Lifelong Learning. 14*(1), 13–30. Retrieved from https://journals.hioa.no/index.php/seminar/article/view/2579

Lawson, M., & Lawson, H. (2013). New conceptual frameworks for student engagement research, policy, and practice. *Review of Educational Research, 83*, 432–479. https://doi.org/10.3102/0034654313480891

Lehtonen, A. (2015). Evaluating students' agency and development of ownership in a collaborative playmaking project. *The European Journal of Social & Behavioural Sciences, 14*(3), 1885–1900. https://doi.org/10.15405/ejsbs.171

Lekies, K. S., Lost, G., & Rode, J. (2015). Urban youth's experiences of nature: Implications for outdoor adventure education. *Journal of Outdoor Recreation and Tourism, 9*, 1–10. https://doi.org/10.1016/j.jort.2015.03.002

McNaughton, M. J. (2006). Learning from participants' responses in educational drama in the teaching of education for sustainable development. *Research in Drama Education: The Journal of Applied Theatre and Performance, 11*(1), 19–41. https://doi.org/10.1080/13569780500437572

McNaughton M. J. (2010). Educational drama in education for sustainable development: Ecopedagogy in action. *Pedagogy Culture & Society, 18*(3), 289–308. https://doi.org/10.1080/14681366.2010.505460

Nabhan, G. P., & Trimble, S. (1994). *The geography of childhood: Why children need wild places.* Beacon Press.

Ødegaard, M. (2001). *The drama of science education. How public understanding of biotechnology and drama as a learning activity may enhance a critical and inclusive science education* (Doctoral dissertation), University of Oslo.

Ødegaard, M. (2003). Dramatic science: A critical review of drama in science education. *Studies in Science Education, 39*, 75–102.

OECD. (2017). Behavioural insights in public policy: Key messages and summary from OECD international events, May 2017. Paris, France: OECD. Retrieved from www.oecd.org/gov/regulatory-policy/OECD-events-behavioural-insights-summary-may-2017.pdf

Ofsted. (2008). *Learning outside the classroom.* Retrieved from www.Ofsted.gov.uk

Özsoy, N., & Özsoy, S. (2018). Creative drama and example of activity plan in stem. *International Education Studies, 4*(4), 213–222. https://doi.org/10.5281/zenodo.1210590

Pajari, K., & Harmoinen, S. (2019). Teachers' perceptions of consumer education in primary schools in Finland. *Discourse and Communication for Sustainable Education, 10*(2), 72–88. https://doi.org/10.2478/dcse-2019-0019

Parzhetskaya, L. (2014). *Enhancing collaborative learning by means of collaborative serious games: providing requirements to collaborative serious games' design* (Master's thesis). University of Oulu. Department of Educational Sciences and Teacher Education. Retrieved from http://jultika.oulu.fi/Record/nbnfioulu-201409201877

Payne, L. (2017). Student engagement: Three models for its investigation. *Journal of Further and Higher Education, 43*, 5. https://doi.org/10.1080/0309877X.2017.1391186

Proyer, R. T., Gander, F., Bertenshaw, E. J., & Brauer, K. (2018). The positive relationships of playfulness with indicators of health, activity, and physical fitness. *Frontier Psychology, 9*, 1440. https://doi.org/10.3389/fpsyg.2018.01440

Reif, N., & Grant, L. (2010). Culturally responsive classrooms through art integration. *Journal of Praxis in Multicultural Education, 5*(1), 100–115.

Rasmussen, B., & Khachik, S. (2000). Mye På Spill (a lot at stake) role-playing and student support: A challenge to both arts and education. *Youth Theatre Journal, 14*, 52–63.

Renninger, K. A., & Bachrach, J. E. (2015). Studying triggers for interest and engagement using observational methods. *Educational Psychologist, 50*(1), 58–69. https://doi.org/10.1080/00461520.2014.999920

Resnick, M. (2018). *Lifelong kindergarten. Cultivating creativity through projects, passion, peers and play*. Cambridge, MA: MIT Press.

Roberts D, & Ousey, K. (2004). Problem based learning: Developing the triggers. Experiences from a first wave site. *Nurse Education* Practice, *4*(3), 154–158. https://doi.org/10.1016/S1471-5953(03)00073-8. PMID: 19038152.

Romero, M., Usart, M., & Ott, M. (2015). Can serious games contribute to developing and sustaining 21st century skills? *Games and Culture, 10*(2), 148–177.

Siklander, P., Ernst, J., & Storli, R. (2020). Young children's perspectives regarding rough and tumble play: A systematic review. *Journal of Early Childhood Education Research, 9*(2), 551–572. Retrieved from https://jecer.org/wp-content/uploads/2020/12/Siklander-Ernst-Storli-Issue9-2.pdf

Siklander, P., & Impiö, N. (2018). Common features of expertise in working life: Implications for higher education. *Journal of Further and Higher Education, 43*(9), 1239–1254. https://doi.org/10.1080/0309877X.2018.1471126

Siklander, P., & Kangas, M. (2020). Leikillisyys esi-ja alkuopetuksessa: Miten sytyttää innostus oppimiseen? Teoksessa T. Kyrönlampi, K. Mäkitalo, & M. Uitto (Toim.) *Esi- ja alkuopetuksen käsikirja* (pp. 223–241). PS-kustannus.

Siklander, P., Kangas, M., Ruhalahti, S., & Korva, S. (2017, March 6–8). *Exploring triggers for arousing interest in the online learning*. In L. Gómez Chova, A. López Martínez, & I. Candel Torres (Eds.), *INTED2017 proceedings. 11th international technology, education and development conference* (pp. 9081–9089). Valencia, Spain.

Skinner, E. A., Chi, U., & The Learning Gardens Educational Assessment group 1. (2012). Intrinsic motivation and engagement as "active ingredients" in garden-based education: Examining models and measures derived from self-determination theory. *Journal of Environmental Education, 43(1)*, 16–36. https://doi.org/10.1080/00958964.2011.596856

Sousa, D. A., & Pilecki, T. (2013). *From STEM to STEAM: Using brain-compatible strategies to integrate the arts*. Sage Publications.

Sun, L., Siklander, P., & Ruokamo, H. (2018). How to trigger students' interest in ICT-based environments: A systematic literature review. *SeminarNet, International Journal of Media, Technology, and Lifelong Learning, 14*(1), 62–84. Retrieved from https://journals.hioa.no/index.php/seminar/article/view/2597

Toonders, W., Verhoeff, R. P., & Zwart, H. (2016). Performing the future. On the use of drama in philosophy courses for science students.*Science & Education*. https://doi.org/10.1007/s11191-016-9853-3.

Veerporten, D., Westera, W., & Specht, M. (2012). Using reflection triggers while learning in an online course. *British Journal of Educational Technology, 43*(6), 1030–1040. https://doi.org/10.1111/j.1467-8535.2011.01257.x

Winston, J., & Tandy, M. (2008). Beginning Drama 4–11 (3rd Ed.). David Fulton Publishers. https://doi.org/10.4324/9780203868935

Pirkko Siklander (Ph.D., Adjunct Professor, University Researcher) works at the University of Oulu, Faculty of Education, the Learning, Education, and Technology Research Unit (LET). She is a leader of the international master's degree program, 'Learning, Education, and Technology (LET)'. Her research focuses on play, creativity, playfulness and making in collaborative learning contexts. She is an expert in designing playful learning processes and has defended her doctoral thesis on the affordances of playful learning environments for tutoring, playing and learning.

Sari Harmoinen (M.Sc., Ph.D., Adjunct Professor, University Lecturer) works at the University of Oulu, Faculty of Education. She is an Education Dean in the faculty. Her focus is science and math didactics. Her research focuses on interactions, STEM, sustainability and assessment. She is an expert in integration between different context and perceptions for phenomenon. She has defended her doctoral thesis on physics learning.

Transdisciplinarity: Science and Drama Education Developing Teachers for the Future

Jo Raphael and Peta J White

Abstract The following chapter describes what can happen when teacher educators collaborate: a drama pedagogue and a science/sustainability educator. Working together to combine science and drama produced a learning experience for their students that enabled them to experience the ethical dimensions of controversial issues while also learning the science embedded in them. This is considered a transdisciplinary practice as the learning resulted from the combination of both science and drama to produce a space for learning beyond either and both disciplines. Jo and Peta, colleagues and teacher educators applied self-study methodology to explore their practice and resulting student learnings. These teaching and learning practices, by transcending disciplinary boundaries, engaged students and enabled learning and insights difficult to achieve through monodisciplinary approaches. The process can be characterised by the metaphor of symbiosis (commensalism or mutualism). These transdisciplinary pedagogies that embed science understandings through drama strategies are advantageous for pre-service teachers. When these future teachers can experience these practices in a setting where collaboration is modelled and valued, they report a positive influence on their future practices.

Keywords Science education · Drama education · Interdisciplinarity and Transdisciplinarity · The future of work · Sustainability

The Context in Which We Educate

An awareness of the need to prepare pre-service teachers to teach a generation of learners who will require a broad range of knowledge, skills and dispositions for a rapidly changing world is at the heart of this project. The twenty-first century already promises unprecedented environmental challenges, societal change, and an

J. Raphael (✉) · P. J White
School of Education, Deakin University, Burwood, VIC, Australia
e-mail: jo.raphael@deakin.edu.au

P. J White
e-mail: peta.white@deakin.edu au

P. J White et al. (eds.), *Science and Drama: Contemporary and Creative Approaches to Teaching and Learning*, https://doi.org/10.1007/978-3-030-84401-1_9

acceleration of innovation in science and technology. Society requires empathetic and well-informed community members, who are interested in and understand science, are able to think critically, and will challenge perspectives that have more to do with power and short-term financial gains, than the future of the planet and sustaining life. As we consider the future of work we see an increasing reliance on generic skills such as collaboration, creative and critical thinking, and problem-solving skills (Tytler et al., 2018). Increasingly, collaborative skills and experience at working efficiently in interdisciplinary teams will result in higher employability. In the future of work we will market our many skills to a variety of engagements rather than signing up for a single career (Tytler et al., 2018). In response to this need, there are benefits of learning through interdisciplinarity where mutual learning gains come from within each discipline, the interactions between disciplines, as well as what can be learned through the creation of something new–by combining disciplinary knowledges (transdisciplinary learning). Our education systems will change to reflect these new ways of working as future employers demand interdisciplinary teams for their future workforce and societies require well-informed and active citizens.

In this chapter we explore the need in education to re-purpose the disciplinary silos toward collaborative, interdisciplinary learning experiences, through a focus on initial teacher education approaches that will meet the challenges of a fast-changing future. Educators often carry the burden of societal change through the addition of programs that influence and inform young people. This is different. We are proposing that the work we have done across and within the disciplinary boundaries of drama and science education will be the way of the future in how we prepare our future teachers to teach the adults of the future, building those skills and capabilities required in the twenty-first century.

Since the dawn of humankind, the arts of drama and theatre have been ways in which people have made sense of their world. Through drama and theatre humans have played out complex stories that involve understanding a wide range of perspectives, and that consider what it means to live ethically in the world. This chapter describes how the application of drama (pedagogy) was essential to exploring controversial and ethical issues, mostly that have arisen from contemporary science practices and research. The learning of science was possible through the lens of science education; however, the deeper more socially contextualised learning required a focus on drama pedagogy and practice.

The Focus of Our Research

Our chapter also explores the relatively unusual experience of opening the doors of our classrooms to colleagues. We explore our collaboration as two teacher educators, one with expertise in drama (Jo) and the other with expertise in science (Peta). It began as a smaller project inside a larger project investigating our practice in teacher education (White et al., 2020), a project in which eight university colleagues across six different discipline areas came together to apply self-study of teacher education methodologies. Peta and Jo recognised that while we had different discipline areas of expertise, we also held interests in each other's discipline area, and brought this interest into our teaching practice. We wanted to understand more about what might be possible when bringing together science and drama. We wondered what we were missing, and what exemplary, interdisciplinary science in drama and drama in science might look like. We wanted to apply drama pedagogy and practice to get to the socially-linked parts of science otherwise largely unexplored in a traditional science curriculum and context. We also wanted to build the capacity of our pre-service teachers more generally, and to have them develop the social interest informed through scientific literacies that is required of active citizens.

We recognise that we (teacher educators) often teach the same students within course cohorts, yet we often don't seem to collaborate or engage beyond our disciplinary boundaries. We expect our students to draw the connections between the different disciplines and pedagogical approaches when they experience each discipline in isolation. The discipline silos persist even though we know that after graduation, many may teach as primary teachers integrating disciplines through their lessons, or in secondary STEM centres also privileging interdisciplinary ways of working.

It is our experience that in the university context academics are often valued for their specialty expertise, and discipline areas tend to be siloed, such that there is a lack of awareness of what was happening in our colleagues' workshops, seminars or tutorials. Palmer (2012) describes this unfortunate situation as common in many education contexts: "When we walk into our workplace, the classroom, we close the door on our colleagues. When we emerge, we rarely talk about what happened or what needs to happen next, for we have no shared experience to talk about" (p. 170). For this project, we opened the doors of our classroom to create a shared experience, and continuing conversation.

Palmer suggests what we need is "[g]ood talk about good teaching … to enhance both our professional practice and the self-hood from which it comes" (Palmer, 2012, p. 172). In this project we were hoping for honest dialogue—a productive and expansive conversation about teaching practice. We were seeking conversations about teaching teachers that offered an 'opening up', not a 'closing in on', in order to better understand ourselves, our teaching, and our students' learning. In collaborating across our two disciplines we also aimed to build knowledge of arts-based pedagogies, and drama in particular, in the context of education for sustainability and science education. However, we also considered how such collaboration might re-energise our teaching, and create new teaching ideas that would sustain us, as teacher educators who often work in isolation.

We saw value in modelling the practice of working collaboratively as colleagues for the pre-service teachers that we teach. We are demonstrating that we are learning too. Greene (1978) explains that students "are most likely to be stirred to learn when they are challenged by teachers who themselves are learning, who are breaking with what they have too easily taken for granted" (p. 51).

The questions for our project emerged from a list of 'wonderings' (Samaras & Freese, 2006). We asked:

- What happens when we open the doors of our classroom and cross our discipline boundaries?
- What can we learn when our critical friend brings expertise from an entirely different discipline area?
- What connections might we make across boundaries and how do we move from observation and reflection to further action?
- What impressions might our self-study practice and sharing of the classroom space have on our students and our colleagues?

The Curriculum Imperative

As we have already mentioned, we are living in a time of rapid change, characterised as bringing complex environmental, scientific and technological challenges. Finding potential solutions to complex issues requires untangling and understanding the multiple perspectives that stakeholders bring to the issue. As we embarked on this project, we were aware of other studies that reported on the contributions that drama can make to science learning, including learning about concepts, the nature of science, and science's interactions with society (Ødegaard, 2003). It is the area of socio-scientific learning that we found most interesting to explore in our collaboration in science/drama. International studies suggest that many science teachers are reluctant to address controversial socio-scientific issues in their classrooms and would be interested in learning about new pedagogical approaches and models (Saunders & Rennie, 2013). In particular, science teachers do not often adequately utilise the opportunity for accessing human emotions that art integration, and drama in particular, can provide (Turkka et al., 2017, p. 14).

International studies have demonstrated how drama can assist students to think like a scientist, including thinking about complex and controversial issues (Ødegaard, 2003), and the societal influences of the work of scientists (McGregor et al., 2019). Andersen (2004) describes the benefits of drama for providing 'as if' authentic contexts, resembling the professional world of the scientist. Inhabiting an 'as if' science world can assist in developing the necessary dispositions and attitudes of the scientist, such as curiosity and scepticism—'habits of mind' that are difficult to teach out of context (Andersen, 2004, p. 283). In other studies, drama is valued for its potential to assist in the development of argumentation and decision-making skills (Belova et al., 2015), and for increasing scientific dialogue and discourse (Dorion, 2009). Well-prepared drama activity can provide a student-centered social-constructivist approach to learning through problem-solving (Çokadar & Yilmaz, 2010) and can help to provide relevance for students by assisting them in understanding how scientific issues impact upon their lives (Dorion, 2009).

The two classrooms we explore and report on in this chapter involve learning about human interactions and pluralistic views and aim to prepare pre-service teachers to teach across four intersecting areas of the Australian Curriculum: Science, to explore The Arts (Drama), two Cross-curriculum Priority (CCP) areas (Sustainability and Aboriginal and Torres Strait Islander Histories and Cultures), and General Capabilities (Ethical and Creative and Critical Thinking). The intersections between these curriculum areas excite and energise us in this project as the potential for interesting learning adventures is rich.

The curriculum rationale for Science states that: "Science is a dynamic, collaborative and creative human endeavour arising from our desire to make sense of our world through exploring the unknown, investigating universal mysteries, making predictions and solving problems" (ACARA: Science as Human Endeavour, 2020). Providing opportunities for students to develop understanding of their world and the perspectives of others, is also in keeping with the Australian Curriculum rationale for Drama which describes exploration of worlds and understanding different worldviews through roles and characters: "Drama enables students to imagine and participate in exploration of their worlds, individually and collaboratively. Students actively use body, gesture, movement, voice and language, taking on roles to explore and depict real and imagined worlds… and analyse their own and others' stories and points of view" (ACARA: Drama, 2020).

Sustainability is one Cross-curriculum Priority (CCP) that is the focus of our first classroom example in which drama and environmental science are brought together to explore "environmental, social, cultural and economic systems and their interdependence" and the role science plays in providing "the basis for decision-making in many areas of society and that these decisions can impact on the Earth system" (ACARA: Sustainability, 2020). The CCP of Aboriginal and Torres Strait Islander Histories and Cultures is of relevance to our first classroom example that is focussed on the Daintree Rainforest. Students develop awareness of the traditional custodians of the land, the Kuku Yalanji people, and their special connection to Country/Place. They learn how the rainforest has been sustainably managed over thousands of years through Indigenous culture, knowledge and close understanding of country, including the scientific understandings that Indigenous Knowledge contributes to contemporary science (ACARA: Aboriginal and Torres Strait Islander Histories and Cultures, 2020).

The Australian Curriculum highlights the need for learning in the General Capabilities (ACARA: General Capabilities, 2020). A particular area of the Australian Curriculum that we are interested in is learning the capability of Ethical Thinking. This is included in the Australian Curriculum and assessed at all levels of schooling with the aim to "assist students to engage with the more complex issues that they are likely to encounter in the future, and to navigate a world of competing values, rights, interests and norms" (ACARA: General Capabilities, 2020). Also assessed is the capability of Critical and Creative thinking. This requires students to develop skills to think "broadly and deeply using skills, behaviours and dispositions such as reason, logic, resourcefulness, imagination and innovation in all learning areas at school and in their lives beyond school" (ACARA: Critical and Creative Thinking, 2021).

These capabilities are active ways of knowing and being, and include skills, behaviours and dispositions, aimed to assist students to become lifelong learners who "are creative, innovative and resourceful, and are able to solve problems in ways that draw upon a range of learning areas and disciplines and deep content knowledge" (Council of Australian Governments—Education Council, 2019, p. 7). Creativity and innovation are traits valued in both artists and scientists who require the "courage to imagine the unimaginable" (Turkka et al., 2017, p. 1).

It is usual for pre-service teachers to be provided with learning experiences that attend and focus on one of the learning areas while exploring them through the CCP and General Capabilities; however, we were explicitly attempting to broaden these learning experiences to acknowledge a more authentic or real-life exploration.

Methodology

Jo and Peta's research collaboration was part of a larger research project with a methodological focus on self-study (White et al., 2020). The *Collaborative Reflective Experience and Practice in Education* (CREPE) Faculty Research Group was formed in 2014 where a team of eight colleagues worked together in various ways to establish a research collective with several smaller, focused research projects enabling opportunities to apply self-study methodology to specific projects. Ours was the exploration of sustainability education through drama pedagogy–initially. As we opened our classroom doors to each other we realised that additional opportunities were presented to us and we broadened our exploration considerably.

The application of self-study methodology was purposeful as it enabled rich exploration of our research question: *How can we continue to develop our teaching practice to ensure we are high quality, contemporary teacher educators, and practice informed researchers?* This question drew us all together with the desire to develop our teaching and research practice simultaneously and collegially (Tidwell et al., 2009). Self-study methodology has been applied to teacher education research in many ways (Loughran et al., 2007), but it was the collegiality made possible through research scholarship of teaching practices that drew each of us in (Hannigan et al., 2016; White et al., 2020). Samaras (2011) offered a framework that resonated with the research group. She suggests that self-study: *is a personally situated inquiry*—drawing on our own teaching experiences and disciplinary focus to engage in research to investigate our teaching scholarship; *requires critical and collaborative inquiry*: developing a community of practice that supports the continued development of our teaching scholarship; *aims to improve learning*: through our scholarship and professional/personal lives; *employs a transparent and systematic research process:* with critical friend collaborations situated within the overarching project; and *contributes to our fields through knowledge generation and dissemination* (Samaras, 2011, p. 10). Alongside the self-study methodology we operated as critical friends (Bullough & Pinnegar, 2007; Handal, 1999) to "prevent self-deception" (Lomax, 1991, p. 14) and see our practice through the lens of others (Samaras, 2011).

We applied arts-based strategies (Barone & Eisner, 2012; Finley, 2011) to engage and enable deeper reflection and exploration of our teaching scholarship. Much of this work related to collaborative and individual representation as opposed to more dramatic performance or representation so we won't explore these practices in this chapter. These practices have been reported on in other publications (Raphael et al., 2016; White et al., 2020).

It All Started with an Invitation to Explore Education for Sustainability

While the overarching project saw us meeting regularly, utilising research retreats, and planning seminars and learning opportunities for colleagues and students, it was the focussed research project, where Jo and Peta co-designed their own investigations, that saw this drama-science project develop. We used dialogic written reflection on practice as a way of generating data (Berry & Crowe, 2009). Jo began by writing an invitation to Peta to explore how she was attending to sustainability in her Drama seminar. Peta responded in writing, took up the challenge, and participated with the pre-service teachers in the seminars. The process is described in the following sections as we explore the findings.

Opening the Doors of Our Classrooms and Building Disciplinary Connections

This is a collaboration with two parts. The first was an invitation to the science educator (Peta) to join in the class of the drama educator (Jo), drawing on drama to teach environmental science and sustainability for primary pre-service teachers to take into primary classrooms. The second was an invitation to the drama educator to collaborate with the science educator on teaching controversial issues in senior secondary science for pre-service biology teachers to take into secondary classrooms. We present the experience/findings through the open doors of two classrooms.

Classroom 1—Drama for Sustainability

Jo's primary field of teaching is drama and theatre with a focus on drama applied to learning or drama pedagogy. She teaches teachers to teach drama as an art form within the Arts curriculum, as well as introducing them to ways to draw upon drama pedagogy to teach other subject areas. As part of this work, Jo designed workshops on teaching concepts of environmental studies and sustainability through drama (Raphael, 2015). As previously described, the Australian Curriculum places emphasis on Sustainability as one of the priority areas for study that connects and relates relevant aspects of content across learning areas and subjects (ACARA: Sustainability, 2020).

Jo asked Peta to join her workshop to prepare pre-service teachers to teach the cross-curriculum priority area of sustainability, and to experience Jo's class in a similar way to how the pre-service teachers do, without any additional preparation or prior knowledge. For Jo, this was a workshop that she had taught in primary schools as well as for pre-service teachers refined over time through reflection on her practice. However, inviting Peta, who brought an environmental scientist's eye, might offer new perspectives, casting new light on aspects of Jo's teaching, potentially prompting Jo to notice things that may have been ignored or taken for granted (Russell, 2010).

Drama for Sustainability—Part 1

The basis for the Drama for Sustainability workshop was Jeannie Baker's picture-story book *Where the Forest Meets the Sea* (1987) with detailed three-dimensional collages as illustrations of the tropical Daintree Rainforest of far-north Queensland. In this story we follow a young boy, who is full of wonder as he walks through the rainforest observing and reflecting on the natural environment. Before reading the text, students are engaged in activities designed to discover their brought knowledge and build greater awareness of the rainforest ecosystem. This is done through talking, reading, and looking at images, maps and artefacts. An episodic approach to drama (Ewing & Saunders, 2016) allows for different learning activities to be inserted where most relevant in a staged reading of the book, including whole class drama activities such as making the soundscape of the rainforest, imagining and miming walking through the rainforest environment. In a central activity, the students are invited to create the rainforest in the classroom by choosing one living or non-living element to embody in the space in relation to other elements. They enter one-by-one: "I am a tall tree reaching toward the sun"; "I am a fern growing in the shade of the tree"; "I am a cassowary searching for insects under the fern". When the complex rainforest has been formed, we all recognise the interdependence of the elements of the ecosystem–a concept that is all the more profoundly understood when the teacher, in the role of forester, enters the rainforest scene and fells and removes the largest trees. One-by-one, each of the elements impacted describes their fate and removes themselves from the forest. Usually groaning with disappointment, they explain the effect, for example, "I'm a fern and I'm shrivelling up because I have lost my shade". Typically, we are left with the sun beating down on rocks, a muddy river, a drying forest floor and possibly a sapling enjoying increased sunlight and promising hope.

A Closer Look at the Learning in Drama for Sustainability—Part 1

Inviting the students to take on the role of the natural elements of the rainforest, either living or non-living, serves to build their knowledge and understanding of the complexity of the rainforest ecosystem. However, when it comes to environmental sustainability, knowledge is not enough to create a lasting change in personal practice, what is needed is an engagement of the affective (Everett et al., 2009). When the trees are removed and students in role find their being is impacted, they have a visceral response, it is as if they feel the pain themselves. Suddenly they understand that the removal of the trees matters, and they gain a sense of empathy for the living things in the rainforest. The imaginative world of drama allows for a shift from an anthropocentric worldview, providing a feeling of empathy with other forms of life on planet Earth. As the rainforest disassembles as a result of the removal of the biggest trees, the students become aware of the interdependence of the elements of the ecosystem. The pre-service teachers discuss how ecosystems might be taught by other approaches such as diagrams, discussion, video, or even walking in the environment. However, the power of the drama approach is that the students have actually been a part of the ecosystem, they embodied it, saw it, heard it and even felt it, and they do this collectively. The discussion that followed was rich and inclusive of many voices.

Drama for Sustainability—Part 2

Through additional information and artefacts, the students learn about the Kuku Yalanji people, the Indigenous custodians of the Daintree Rainforest, and their stewardship of country for millennia. The students learn Indigenous names for rainforest plants and their properties and uses, as well as Kuku Yalanji customs that have ensured sustainable management of the rainforest. They share this knowledge through drawing symbols and oral storytelling, adopting drama for learning in an Aboriginal way (Marshall, 2004).

After a range of drama activities related to the text, students begin to feel a close affinity to the rainforest. and they are like the boy in the book who wistfully wonders "will the forest still be here when we come back?". In the final pages of the book, Baker's collage of the natural rainforest and coast is superimposed with the signs of a tourist resort, hotels, roads, towers and a marina. The students are invited to imagine a newspaper headline: "Developers face divide over rainforest resort plan", we determine the rainforest resort has not yet been built, but there is pressure to do so. Students think about the stakeholders in this proposal and make a list that includes local residents, local business, Kuku Yalanji people, scientists, environmentalists, tourists, developers and members of government. Students are assigned roles randomly as representatives of one stakeholder group. They are encouraged to discuss the range of views (nuanced, rather than purely stereotypical) that individual stakeholders might hold within these groups. A whole class role-play is conducted in which the teacher, in the role of convenor of a town hall meeting, facilitates a discussion about the proposed development and each stakeholder has a chance to speak about their point of view.

A Closer Look at the Learning in Drama for Sustainability—Part 2

Aboriginal ways of knowing are focussed on relationships as well as embodied and oral traditions and have much in common with learning through drama (Marshall, 2004). The affective experience of these traditions allows for an understanding of the special connection to Country/Place, both physical and spiritual, experienced by Australia's First Nations Peoples. In this workshop, once the students have become aware of the interdependence of the rainforest ecosystem, seeing the rainforest from a Kuku Yalanji perspective where each element may be food, medicine, shelter or of spiritual significance, highlights the importance of maintaining the balance to sustain the rainforest, which in turn sustains human life. Students can begin to understand that the Kuku Yalanji have employed scientific knowledge and practices that have sustained the ecosystem for millennia and have much to contribute to contemporary science.

Throughout the drama workshop the students develop a sense of close connection to the rainforest. When the final pages of the book are revealed, and students see the rainforest may be lost if developed into a luxury resort, we sense they experience a sudden shock. There is a tendency at this point for students to unequivocally argue for no change to the natural environment, seeing any level of development as negative. However, we remind students that humans often do have compelling reasons to develop land, and that we enjoy living in built environments that have required some destruction of natural habitats. The problem is, how can we do this is a sustainable way? The final activity that brings stakeholders together for a whole class role-play, is modelled in one session for the pre-service teachers, but is best done over a longer timeframe, when students have time to research and prepare their roles, as in a more thorough employment of the Mantle of the Expert approach (O'Neill, 2014). However, the result is the same, a lively and sophisticated debate, where the problem is viewed from many perspectives and its complexity is appreciated. If problems are to be solved, then the complexity needs to be understood, and the students work hard in the drama role-play to explain, understand, question, challenge, persuade and imagine possibilities as they attempt to find a solution. This is followed by crucial discussion out of role, to take stock of the learning, and conduct a reality check. Before the conclusion, we always explain that in the time since Baker wrote *Where the Forest Meets the Sea*, debate and activism has led to conservation and protection of the Daintree Rainforest, including a World Heritage listing. Having a teacher with an environmental science focus helped to facilitate the post-workshop discussion, to make connections with the real world, and move into futures thinking. This includes what they can do personally regarding environmentally sustainable practices, and as teachers with their future students.

Classroom 2—Teaching Controversial Issues in Secondary Science

Observing these classes prompted Peta to wonder how some of the drama pedagogies used in Jo's workshops might be adapted to the teaching of controversial issues in her senior secondary biology class, which was running concurrently that trimester. The self-study project then took a practical turn as we collaborated to deliver a workshop that employed drama as a pedagogy to explore the complexity within a contemporary scientific topic: stem cell research. Our aim was to build understanding amongst the pre-service teachers about controversial issues in scientific research and practice, to use drama to build understanding and appreciation of the diversity of views in this complex issue, and to talk to the pre-service teachers' about how to unpack or explore other controversial issues in senior secondary biology.

The introduction of these drama activities within the science laboratory space was intriguing and a little unsettling for the pre-service biology teachers. To ease them in, Jo began with a short presentation about drama as pedagogy, with reference to thinking frameworks and tools (Moseley, 2005), such as taking other people's perspectives. She provided examples of dramatic role-play and simulations for secondary students in different contexts such as the Model United Nations, and the Old Melbourne Gaol Crime and Justice Experience court room drama. This introduction served to reassure the students that the drama experience they were about to engage in was an example of established pedagogical practice and offered as a model for exploring complex issues.

In order to prepare the drama experience, Jo had to conduct some research to understand the controversial issues surrounding stem cell research. This preparation is an important and necessary step for the drama teacher; it requires spending time sourcing and reading scientific information and commentary on the issue. While Peta already had some familiarity with the science of stem cells, Jo took the opportunity to study the science. Together we worked out which issues would be most interesting to explore, and in our first iteration of the workshop, we focused on the controversial use of embryonic stem cells. Scientific understandings are constantly developing, so in later versions of the workshop, when pluripotent stem cells reduced the need for embryonic stem cell use, our focus issue became the ethics of stem cell tourism.

After the topic of stem cells was unpacked, the pre-service teachers were given some time to conduct some research about the controversies and issues in stem cell science using iPads. They shared what they had learned making a collaborative list of the various people/groups who were stakeholders in the issue. They listed stakeholders including patients, family members, medical doctors, stem cell researchers, stem cell therapy providers, government officials, religious leaders, media and others. From this list, we selected some key stakeholder groups and randomly distributed one role to each group of two or three students. Within these groups, the students discussed the various views that may be held by the stakeholder group they had been allocated. Our aim was for them to move beyond considering only stereotypical views, to more nuanced views that individuals within the different stakeholder groups might have.

The context of the whole class role-play was given, and we chose the situation of a television debate program in which a host (teacher in-role) holds a discussion with a panel of experts and representative stakeholders. The students prepare to take on their roles which may include finding a signifier, such as a badge, costume or prop to help them play that role. The science room is transformed into the television studio, and the teacher, in-role as host, briefs the panel members before going 'live to air'. A lively debate is held, with the host questioning strategically to ensure a wide range of contrasting views are heard.

The drama activity is followed by two layers of critical reflection and discussion. The first is in response to the learning about the controversial issue, and when necessary science content is corrected, clarified or added. Because these are pre-service teachers, the second layer of discussion is about the affordances of the drama as pedagogy and what they learned through this process.

A Closer Look at the Learning in Teaching Controversial Issues

Being positioned as a teacher/leader in the middle of a controversial issue can be challenging. Everyone has opinions, some work to become better informed (as modelled by Jo), others rely on social discourse to provide their opinions. It is inevitable that many science teachers will experience scenarios where the community opinion is biased on an environmental issue with social implications. Learning strategies that enable student agency and empowerment, as well as critical and creative thinking in the development of scientific literacies become valuable. Drama pedagogies that enable students to explore ideas and opinions that might be dislocated from their own is imperative. The fiction of the drama, and the taking on of roles, allows for setting aside personal beliefs in order to explore a wider range of perspectives and opinions on the science. In fact, some of the insight needed to understand and appreciate the science involved in these controversial issues cannot be easily accessed and this is why drama pedagogies that enable emotional connections, affective learning, and student agency are used to great effect. Without the use of drama pedagogies, the deep understanding of the human view in the socio-scientific issue could not be achieved.

This scenario developed further. Jo and Peta offered a similar workshop at a science teacher conference. They were amazed at how many participants signed up for their session (over 100). They had to adapt the strategy to cater for the large volume of participants but managed to facilitate a useful workshop where many teachers experienced the agency as well as learned new insights about the science research and practices. The scenario then further developed into a project for 'Reconceptualising Mathematics and Science Teacher Education Program (ReMSTEP)' where drama pedagogy became central in scaffolding how students could engage with the controversial issues surrounding stem cell research (White et al., 2018). A leading research scientist collaborated to design learning materials (video) that could inspire and engage students in not just the science but the idea of being a scientist. Peta and Jo have written about this in another chapter (White & Raphael, In Press).

Modelling Collaborative Practice for Pre-Service Teachers

Self-study as a methodology is at its most powerful, according to Russell (2010), "when a teacher educator is willing to explore the complexities of learning by those who wish to become teachers in the context of exploring simultaneously the complexities of one's own learning to teach" (p. 691). We discovered that bringing together the disciplines of drama and science had positive impacts on our own teaching practice. It also had important influences on our students who became interested in the approach.

> When looking at the self-study that Peta and Jo are a part of, I am extremely interested. I think it is incredible that even at high levels of education (university level) there is room for growth and for evaluation of the teaching. (Pre-service teacher 1)

The presence of another educator from a different, but complementary discipline area, appeared to encourage deeper thinking about pedagogy amongst the pre-service teachers.

> It seems more productive in a way to get the opinion of someone who operates in contrast to you as it would provide an insight from another type of educator and thinker. ... sometimes it is easy to find a problem with your own work but more difficult to solve it. I'd definitely like to use this technique and can see it being extremely valuable in my early career. (Pre-service teacher 2)

We also recognised interest and respect amongst the pre-service teachers for the way we modelled collaboration and reflective practice as educators with a view to improving practice. Some saw it as an invitation.

> This process allows me to have questions and curiosities about my practice. And it is these curiosities and questions that empower higher order thinking and a stronger will to achieve higher standards. (Pre-service teacher 3)

Our presence in each other's teaching spaces helped to generate rich and authentic discussions about teaching and learning that positioned us all (teacher educators and pre-service teachers) as teachers and learners together (Freire, 1970). It was seen as a way of sustaining and empowering us to innovate during our careers as critically reflective practitioners exploring the interplay between different discipline areas.

> This environment will help teachers stay motivated, passionate, current with their knowledge and most importantly[ly] stimulates a critical environment where teachers feel safe to challenge the status quo of education. (Pre-service teacher 4)

Jo and Peta often reflect on what it meant to 'open their classrooms' to each other. The power of combining not only experience but disciplinary expertise was evident as the collaboration deepened. Peta reports learning how to use 'teacher in role' as a strategy to engage all of her students simultaneously. She also reflected that Jo invested in learning the science, making the time to understand stem cell research so that she could perform accurate detail with and with clarity. Jo felt energised by the opportunity to bring drama pedagogy to challenging contemporary socio-scientific issues and enjoyed learning about them in the process. Her appreciation of the potential of drama was reinvigorated when the drama approach to science learning was enthusiastically received by her science colleague and pre-service teachers. This set her on a new quest to understand how best to equip and support pre-service teachers from a non-drama background to effectively take up this approach in their own teaching.

Conclusions

Accessing science learnings that could not otherwise be accessed or enabled in a classroom setting in such a short amount of time is one significant advantage of applying drama pedagogies in a science context. This is a form of interdisciplinarity, as it requires one discipline to support the learning of another. Jo and Peta experimented with appropriate metaphors to express the application of science drama: their favourite one was 'commensalism' (mostly because of the biological reference) (Wikipedia, 2020). This is a specific form of symbiosis where one species (a discipline–such as science) benefits (i.e. greater student learning) while the other species (discipline–drama) is neither harmed nor benefited. The model could be extended to appreciate that 'parasitism' is undesirable as either science or drama (as a discipline) would benefit at the expense of the other (as in students would not learn disciplinary concepts or practices and might even obtain negative practices or beliefs about the discipline). Of course 'parabiosis' means no gain in a learning context as both species (disciplines) might be practiced without interaction. Another form of symbiosis, possibly the most admirable, is 'mutualism' where both science and drama learning take place through the pedagogical and contextual learning situations. To some extent this occurred in Classroom 1, the drama education class for pre-service teachers, when sustainability and the text were explored through multiple dramatic activities, over a longer timeframe, with attention to the aesthetics of the artform. We were more cognisant of learning about drama, artistry and artful learning (Dunn & Stinson, 2011; Østern, 2019) because the focus was in preparing pre-service teachers to teach the art of drama; environmental science provided us with a rich theme to explore and learning in drama and in science was able to occur.

We suggest that our symbiotic pedagogies were, in fact, a transdisciplinary approach as they generated new knowledge as a result of disciplinary interactions and learning adventures not limited to science or drama. Learning in drama extended the science beyond straight knowledge to consider the wider context that is so important, concerning science's impact on people and societies, and science offered an opportunity for extending thinking about the role and impact of drama in wider contexts.

There is an intertwining of the disciplines and at the points where they interconnect, we see the sparks of new learning that would not be possible for one without the other. This approach to learning seems imperative when exploring our many and continuing environmental and social justice challenges. The intent is to bring our students (pre-service teachers in our case) to a new place where they learn disciplinary skills and practices in relevant contexts but with application that is widely relevant to a variety of teaching and learning contexts. It is this practice in learning that will serve our students well for unknown and unknowable futures. Focusing on skill building, collaboration, and personal connection is crucial to this approach.

Acknowledgements This research was carried out with the support of a Faculty of Arts & Education Research Grant, Deakin University.

References

ACARA: Aboriginal and Torres Strait Islander Histories and Cultures. (2020). Retrieved from https://www.australiancurriculum.edu.au/f-10-curriculum/cross-curriculum-priorities/aboriginal-and-torres-strait-islander-histories-and-cultures/

ACARA: Critical and Creative Thinking (2021) Retrieved from https://www.australiancurriculum.edu.au/f-10-curriculum/general-capabilities/critical-and-creative-thinking/

ACARA: Drama (2020) Retrieved from https://www.australiancurriculum.edu.au/f-10-curriculum/the-arts/drama/

ACARA: Science as Human Endeavour (2020) *Structure: The three interrelated strands of science*. Retrieved from https://www.australiancurriculum.edu.au/f-10-curriculum/science/structure/

ACARA: General Capabilities (2020) *Ethical understanding*. Retrieved from https://www.australiancurriculum.edu.au/f-10-curriculum/general-capabilities/ethical-understanding/

ACARA: Sustainability (2020) Retrieved from https://www.australiancurriculum.edu.au/f-10-curriculum/cross-curriculum-priorities/sustainability/

Andersen, C. (2004). Learning in 'as if' worlds: Cognition in drama in education. *Theory into Practice, 43*(4), 281–286.

Baker, J. (1987). *Where the forest meets the sea*. Walker Books.

Barone, T., & Eisner, E. (2012). *Arts based research*. Sage.

Berry, A., & Crowe, A. R. (2009). Many miles and many emails: Using electronic technologies in self-study to think about, refine and reframe practice. In D. L. Tidwell, M. L. Heston, & L. M. Fitzgerald (Eds.), *Research methods for the self-study of practice* (pp. 83–89). Springer.

Bullough, R. V., & Pinnegar, S. E. (2007). Thinking about the thinking about self-study: An analysis of eight chapters. In J. J. Loughran, M. L. Hamilton, V. K. LaBoskey, & T. Russell (Eds.), *International handbook of self-study of teaching and teacher education practices* (pp. 313–342). Springer.

Belova, N., Eilks, I., & Feierabend, T. (2015). The evaluation of role-playing in the context of teaching climate change. *International Journal of Science & Math Education, 1*, 165. https://doi.org/10.1007/s10763-013-9477-x

Çokadar, H., & Yılmaz, G. (2010). Teaching ecosystems and matter cycles with creative drama activities. *Journal of Science Education and Technology, 19*(1), 80–89. https://doi.org/10.1007/s10956-009-9181-3

Council of Australian Governments Education Council. (2019). *Alice springs (Mparntwe) education declaration*. Education Council Secretariat. Available at: https://docs.education.gov.au/documents/alice-springs-mparntwe-education-declaration

Dorion, K. (2009). Science through drama: A multicase exploration of the characteristics of drama activities used in secondary science classrooms. *International Journal of Science Education, 31*(16), 2247–2270.

Dunn, J., & Stinson, M. (2011). Not without the art!! The importance of teacher artistry when applying drama as pedagogy for additional language learning. *Research in Drama Education: THe Journal of Applied Theatre and Performance, 16*(4), 617–633. https://doi.org/10.1080/13569783.2011.617110

Ewing, R., & Saunders, J. N. (2016). *The school drama book: Drama, literature and literacy in the creative classroom.* Currency Press.

Everett, L., Noone, G., Brooks, M., & Littledyke, R. (2009). Education for sustainability in primary creative arts education. In M. Littledyke, N. Taylor, & C. Eames (Eds.), *Education for sustainability in the primary curriculum* (pp. 180–206). Palgrave Macmillan.

Finley, S. (2011). Critical Arts-based Inquiry: The pedagogy and performance of a radical ethical aesthetic. In N. K. Denzin & Y. S. Lincoln (Eds.), *The SAGE handbook of qualitative research* (4th ed., pp. 435–450). SAGE Publications.

Freire, P. (1970). *Pedagogy of the oppressed.* Herder & Herder.

Greene, M. (1978). *Landscapes of learning.* Teachers College Press.

Hannigan, S., Raphael, J., White, P., Bragg, L., & Cripps Clark, J. (2016). Collaborative reflective experiences and practice in education explored through self-study and arts-based research. *Creative Approaches to Research, 9*(1), 84–110.

Handal, G. (1999). Consultation using critical friends. *New Directions for Teaching & Learning, 79*, 59–79.

Lomax, P. (1991). *Managing better schools and colleges: The action research way.* Longdunn Press.

Loughran, J. J., Hamilton, M. L., LaBoskey, V. K., & Russell, T. (2007). *International handbook of self-study of teaching and teacher education practices.* Springer.

Marshall, A. (2004). Singing your own Songlines: Approaches to indigenous drama. In M. Mooney & J. Nicholls (Eds.), *Drama journeys.* Currency Press.

McGregor, D., Baskerville, D., Anderson, D., & Duggan, A. (2019). Examining the use of drama to develop epistemological understanding about the nature of science: A collective case from experience in New Zealand and England. *International Journal of Science Education, Part B, 9*(2), 171–194. https://doi.org/10.1080/21548455.2019.1585994

Moseley, D. (2005). *Frameworks for thinking: A handbook for teaching and learning.* Cambridge University Press.

Ødegaard, M. (2003). Dramatic science. A critical review of drama in science education. *Studies in Science Education, 39*(1), 75–101.

O'Neill, C. (2014). *Dorothy Heathcote on education and drama: Essential writings.* Taylor and Francis.

Østern, A.-L. (2019). Artful teaching of drama-based story-line. In Á. H. Ragnarsdóttir & Björnsson (Eds.),*Drama in education: Exploring key research concepts and effective strategies* (pp. 39–53). Routledge.

Palmer, P. J. (2012). *The courage to teach: Exploring the inner landscape of a teacher's life* (2nd ed.). John Wiley & Sons Inc.

Raphael, J. (2015). Drama and imagination in education for sustainability. In S. Schonmann (Ed.), *International handbook of arts education: Wisdom of the many.* Waxmann.

Raphael, J., Hannigan, S., & White, P. (2016). Drawing out understandings of collaborative self-study in teacher education. In D. Garbett, & A. Ovens. (Eds.), *Enacting self-study as methodology for professional inquiry.* (pp. 109–118). Retrieved from https://www.castleconference.com/conference-history.html

Russell, T. (2010). Self-study by teacher educators. *International Encyclopedia of Education, 3*, 689–694.

Samaras, A. P. (2011). *Self-study teacher research: Improving your practice through collaborative inquiry.* SAGE.

Samaras, A. P., & Freese, A. R. (2006). *Self-study of teaching practices.* Peter Lang.

Saunders, K. J , & Rennie, L. J. (2013). A pedagogical model for ethical inquiry into socioscientific issues in science. *Research in Science Education, 43*(1), 253–274.

Tidwell, D. L., Heston, M. L., & Fitzgerald, L. M. (2009). *Research methods for the self-study of practice*. Springer.

Turkka, J., Haatainen, O., & Aksela, M. (2017). Integrating art into science education: A survey of science teachers' practices. *International Journal of Science Education, 39*(10), 1403.

Tytler, R., Bridgstock, R., White, P., Mather, D., McCandless, T., & Grant-Iramu, M. (2018). *100 jobs of the future: A study commissioned by Ford Australia*. Retrieved from https://100jobsofthefuture.com/

Wikipedia. (2020). *Commensalism*. Retrieved from https://en.wikipedia.org/wiki/Commensalism

White, P. J. & Raphael, J. (Forthcoming). Drama for teaching Controversial Issues in Science. In D. McGregor (Ed.), *Researching learning science through drama: Exploring a range of international perspectives. ESERA Book Series*. Springer.

White, P. J., Raphael, J., Hannigan, S., & Cripps Clark, J. (2020). Collaborative reflective experience and practice in education: A model for a self-study community of practice. *Australian Journal of Teacher Education, 45*(8). https://ro.ecu.edu.au/ajte/vol45/iss8/6/

White, P., Tytler, R., & Palmer, S. (2018). Innovation 5—Exploring models of interaction between scientists and pre-service teachers. In S. Dinham, D. Corrigan, D. Hoxley, & R. Tytler (Eds.), *Reconceptualising maths and science teaching and learning* (pp. 108–135). ACER press.

Jo Raphael (B.Ed, M.Ed., Ph.D, SFHEA) is senior lecturer in drama education at Deakin University. She teaches in postgraduate and undergraduate pre-service teacher education programs and remains active in school and community settings. She has applied drama for learning within diverse contexts including museums and galleries and in areas of the humanities, languages, science and environmental sustainability. Jo is Artistic Director of Fusion Theatre, an inclusive community-based theatre company. She has won multiple awards for her teaching and has been awarded for her extensive contributions to her professional community. Her research and publications span the areas of arts curriculum and pedagogy, teacher education, inclusive education and teaching for diversity.

Peta J White (B.Sc (Hons), Dip Ed., M Rur Sys Man, Ph.D, SFHEA) (B.Sc (Hons), Dip Ed., M Rur Sys Man, Ph.D, SFHEA) is a science and environmental education senior lecturer at Deakin University. Peta has worked in classrooms, as a curriculum consultant and manager, and as a teacher educator in several jurisdictions across Canada and Australia. Peta gained her Ph.D. in Saskatchewan, Canada where she focussed on learning to live sustainably which became a platform from which to educate future teachers. Her passion for initial teacher educator, environmental education/academic activist work, and action-orientated methodologies drives her current teaching/research scholarship. Peta's current research interests follows three directions including: science and biology education; sustainability, climate change, and environmental education; and collaborative/activist research.

Art-Science Education in the Anthropocene: Embodied Metaphor with Puppets and Performance

Shelley Hannigan and Joseph Ferguson

Abstract The Melbourne Zoo's campaign for endangered species (Melbourne Zoos, n.d.), calls for the public to learn about and act on the plight of endangered species. In response to this campaign, researcher-educators designed, implemented and evaluated a sequence of lessons in which art (particularly drama) and science were enmeshed. This learning sequence was designed to support year 10 students from an all-girls school to: (1) learn about endangered species, (2) create puppets of these endangered species using recycled materials, (3) use these puppets in a small mobile 'theatre in a suitcase' performance so students could, (4) communicate the story and plight of their assigned endangered species to a young audience (primary and pre-school children, and their parents). This learning sequence is shared and discussed as a model of teaching and learning art and science education. In doing so, we discuss the power of metaphor that supports young people to connect with science through appreciating the natural world and its species as a 'shared world'; a world with which they are deeply connected. We argue that these learning experiences offer students an opportunity to respond to environmental issues that when done in interdisciplinary and activist ways can potentially address ecological grief and anxiety issues. We share examples from our data analysis to show how the inclusion of arts with science provided opportunities for students to engage with materiality and embodied experiences as they created and 'brought to life' puppets as metaphorical beings in effective, affective and engaging ways.

Keywords Anthropocene · Art-science education · Endangered species · Metaphor · Performance

S. Hannigan (✉)
Faculty of Arts & Education, School of Education, Deakin University, Pigdeons Road, Waurn Ponds, Australia
e-mail: shelley.hannigan@deakin.edu.au

J. Ferguson
Faculty of Arts & Education, School of Education, Deakin University, Burwood, Australia
e-mail: joe.ferguson@deakin.edu.au

P. J White et al. (eds.), *Science and Drama: Contemporary and Creative Approaches to Teaching and Learning*, https://doi.org/10.1007/978-3-030-84401-1_10

Introduction: Standing on the Precipice

We currently stand on the precipice of a climate apocalypse. Human-induced climate change is wreaking havoc with global weather that necessarily leads to a profound disturbance to natural equilibriums. In Australia, where the authors call home and where this research was undertaken, 2020 started with unprecedented bushfires ravaging flora and fauna across the country, leading to the demise of entire populations and the extinction of some species (Mannix, 2020), including the decimation of human communities (Bradley, 2020). The same narrative is unfolding across the globe, as we (humans, the life form solely responsible for this catastrophe) acutely experience ecological anxiety and grief in response to the rapidly approaching death of the planet (Cunsolo & Ellis, 2018; Ellis & Cunsolo, 2018; Oar & Taylor, 2020).

Historically, art and science (as distinct disciplines, but also as a synergy) have been a refuge for us as we come to negotiate our greatest challenges (Sontag, 1966); the case is no different today. Artists (Rosen, 2020) and scientists (Vince, 2020) alike are embracing their particular disciplinary ways of knowing and doing to come to terms with the grief and anxiety that accompanies devastating environmental change. Increasingly, the traditional boundaries between disciplines are dissolving as art and science come together to give us hope that is deeply rooted in the harsh realities that currently confront us–a contemporary case in point is The Climate Museum in New York (2020).

As educators who design, teach and research a range of education programs, we are particularly concerned with young peoples' experiences of ecological grief and anxiety, which is often more acute than that of adults, as Generation Z and Generation Alpha more honestly face up to the apocalyptic nature of a present that they have not themselves constructed, and have their futures pulled from under them by actions from the past not of their doing (Hickman, 2019). In trying to 'be' with integrity in this age of the Anthropocene, the complex and rich interconnection of selves to places (and their organic inhabitants, with their associated inorganic entanglements) is increasingly coming to be known and valued (Ellis & Cunsolo, 2018; Kemmis, 2019). Prompted by those who have the courage to directly engage with the enormity of the situation (see for example the work of the artist/ecologist Daniela Molnar, 2020), we are forced to consider the question: 'Can we reimagine a way to be human that doesn't cause so much suffering for others?' (Rosen, 2020). Drawing on the words of Leslie Head (2016), a geographer hopeful of the possibility of new human–environment relations, we ask: how can we realise ourselves as "Anthropoceneans" (p. 187) in new and yet-to-be imagined ways?

These ideas and concerns informed and motivated for us the design of a teaching, learning and research program in partnership with Melbourne Zoo's campaign for endangered species (Melbourne Zoos, n.d), school-based educators and a group of year 10 students from an all-girls school. We co-designed, co-facilitated and researched this 8-week program of project-based learning (PBL) classes that incorporated art (particularly sculpture/material technology and drama) and science. We were motivated by the potential value of this bringing together of art and science for these young people to productively negotiate these increasingly troubled and troubling times of the Anthropocene, and the associated ecological grief and anxiety, particularly when it comes to the demise of entire species. The program was designed and implemented such that these young people could eventually (in Week 8 of the program) encourage their audience (primary and pre-school children, and their parents), to join them on their ongoing journeys of realising themselves as Anthropoceneans in new and yet-to-be imagined ways.

Our chapter shares and discusses the design of this learning program and some findings from our research in two stages, to show the way in which this project exemplifies how this campaign (and similar initiatives) offers opportunities for students and their educators to open up possibilities of new ways of being human. We share this work in the hope of contributing to considerations of art (drama)/science education programs for the future. We argue that through the material design, construction and performance of puppets as a 'theatre in a suitcase', which is the main focus of this chapter, that students enacted learning as and through the mind/body in a fundamentally non-Cartesian way as they harnessed the metaphoric power of puppets for transformational learning (O'Hare, 2005) and potentially transformational healing (Bernier, 2005). We suggest how various representational practices, in particular that of metaphor, combine with the epistemic power of the body and the materiality of our surrounds, to enable new ways of thinking and doing (in particular for young people) in these rapidly changing times. We propose how embodiment/materiality, in the form of puppetry in the educational context, can form part of the process of 'being human' in new and fundamentally different ways, or, as Hasbach (2015) notes, "to fully experience our totemic selves–that part of us that recognizes our kinship with the more-than-human world and our dependence on it" (p. 205).

Theoretical Framing: The Metaphorical and Material Aliveness of Puppets

Science is increasingly combined with drama in exciting ways for educational purposes–particularly in early childhood and primary classes (Amsen, 2019). 'Team Hyena Puppet' (Trommer-Beardslee et al., 2019) explored the power of puppetry to support student understanding of the ecology of sand dune succession and the threat-sensitive behaviour of spotted hyenas. While Simon et al. (2008) demonstrated the power of puppets to promote student talk conducive to reasoning in the science classroom. Whether through education to develop new understandings (O'Hare, 2005) or therapy to engage with emotions and embrace creativity (Bernier, 2005), the power of puppets as a particular performative practice is emerging as a key component of drama. Puppets are often powerfully present in dramatic performances as metaphorical beings with which we identify (Anderson, 1987; Ackerman, 2005; Gross, 2012; Hausman, 1989; John & Carol, 2011). Ackerman (2005, p. 7) points out how puppets can "become visual metaphors for ideas, characters, or emotions" which for us held a lot of appeal for students to develop their learning of endangered species and communicate this learning through/with puppets. This was particularly important for students when communicating with younger people (primary and pre-school children) as "puppets have a magic and a power when we think of them as metaphors" (O'Hare, 2005, p. 3).

The metaphorical power of puppets has much to do with the embodied and material nature of puppetry performances. We take seriously Mello's (2016, p. 49) proposition that "puppet and material performance is an embodied practice that can be understood as a theory and technique of trans-embodiment". This is not simply a case of actor-puppeteer 'activating' the puppet, but rather a transference 'between' actor-puppeteer and puppet: "a dialogue between live performer and material bodies, each affecting the other" (Mello, 2016, p. 50). It is only through this process that the puppet as a material object can acquire "anima–meaning its soul or vital force–or aliveness" (Mello, 2016, p. 54). It is through both the direct transfer of body language and gesture, and the indirect infusion of emotion and memory from the actor-puppeteer to the puppet, which makes this possible. Through this process, the actor-puppeteer, and indeed the audience-participant, is also transformed in their aliveness as they enter "the umwelt of puppets" in which "puppet-things can exert agency over people" (Cohen, 2007, p. 129). We suggest in this chapter that puppetry can do something similar for how we (as humans, in particular young people who are trying to imagine new possible futures) relate to nonhumans (specifically endangered animals).

When we talk of embodiment, we avoid favouring either the mind or the body. But there is a need, we feel, to foreground the body to push back against the centuries old Cartesian tradition of locating the epicenter of meaning-making in the mind. When it comes to doing science and teaching/learning science: "We do not sit inside our own bodies looking out. We are our bodies, actively scanning and looking for, looking around, reacting to visual impressions, anticipating them, comparing expectation to current sensation" (Lemke, 2015, p. 5). Similarly, art education and the practice of art is (or at least should be) about "the embodied experience of being" (Mathews, 1998, p. 1).

Methodological Considerations: The Matter of Nature and the Nature of Matter

As this research program concerned endangered species as well as creating puppets out of recycled materials and creating mini-portable theatres ('theatre in a suitcase') to perform and communicate, the "matter of nature and the nature of matter" that Cool and Frost (2010, p. 6) highlight was highly relevant. Materiality was an important consideration for the creators of the teaching and learning program, as it introduced students to how "our environment materially and conceptually [is] reconstituted in ways that pose profound and unprecedented normative questions" (Cool & Frost,

2010, p. 6). Unpacking this idea further, this included at least two issues of sustainability: (1) the effects of climate change on endangered animals in our 'backyard', and (2) considering ways to upcycle materials to create puppets for performance. The program was designed to reflect the material nature of doing science (Latour & Woolgar, 1986) and doing art (Berger, 2019); the inherently embodied nature of making and performing (with) puppets was the link between the material and the human. Referring to climate change and the importance of materiality, Cool and Frost (2010, p. 6) noted that "recent developments call upon us to reorient ourselves profoundly in relation to the world, to one another, and to ourselves". As suggested earlier, the materiality of puppetry afforded the opportunity for students in this program to potentially undertake such transformations.

Research Design: What We Did and How We Did It

Our analysis discussed in this chapter draws on data that emerged in Week 3 and Week 8 of the research program, as highlighted in **bold** text in the schedule below (Table 1):

Week 3 was an initial puppet design-creation class which was conducted in a specially built classroom (SLRC) with multiple wall and ceiling-mounted cameras and microphones capable of capturing all individual and group interactions. In

Table 1 Schedule of the 8-week program

Week	Classes, workshops and their locations
Week 1	2.5 h excursion to Melbourne Zoo where students learnt about endangered species of Victoria, Australia
	Students sorted into groups of 3 and each group assigned an endangered species to focus on
Week 2	Two classes in school devoted to students researching their endangered species, in their groups
Week 3	**Monday: 2.5 h class in a specially designed Science of Learning Research Classroom (SLRC) focusing on skeletal systems and puppet construction**
	Friday: Trash-puppet workshop 1 run by the Melbourne Zoo art teacher
Week 4	3 × 1.5 h classes on the project, in school
Week 5	2 × 1.5 h classes on the topic, in school
	Friday: Trash-puppet workshop 2 run by a Melbourne Zoo art teacher
Week 7	3 × 1.5 h classes on the project, in school
Week 8	2 × 1.5 h classes on the project, in school
	Presentation of trash puppets and 'theatre in a suitcase' to primary school children at Melbourne Zoo on National Threatened Species day, 7th September

the first part of this 2.5 h session, a female scientist/textile artist and an artist-art/textile educator introduced students to animal skeletal structures and animal-inspired puppets. This was done to support students to appreciate the link between structure and function that enables particular animals to survive and reproduce in certain environments, which drives evolution (that is necessarily linked to possible extinction). Students were also supported to understand how the materiality of puppets can include complex skeletal structures that can help in their movement, performance, aesthetic make-up, form and function. Toward the end of this first part of the session, students were provided with examples of skeletal structures that they matched to their assigned endangered animal, which they had already started learning about in prior Zoo sessions and classroom sessions. For the second half of this session, students explored the material and creative possibilities of developing puppets using a table piled high with recycled materials, with tools also provided. Photographs were taken of these early puppet construction processes and the final artefacts, which were used as data to support interpretation of the video recordings.

Often before an artist starts creating a sculpture or puppet, drawings are developed and/or maquettes created. The puppet-creations in the Week 3 session were similar in that they were initial '3-dimensional drawings made with materials' that would be developed further (and could even change dramatically) by the students in subsequent sessions. This approach was possible because from this Week 3 session onwards, students would be involved in dedicated trash-puppet making lessons with an art teacher who had previously conducted trash-puppet making workshops. Students would also be attending other lessons that focused on: researching their endangered animal and associated habitat; developing scripts to communicate the plight of their endangered species to younger audiences (as part of supporting the Zoo's campaign); and developing puppets, props and sets for their 'theatre in a suitcase'.

In the final session, Week 8, each student group performed with their trash puppets as they enacted their 'theatre in a suitcase' to communicate the plight of their assigned endangered animal to younger children and their parents at Melbourne Zoo. These performances were filmed, and students were interviewed (audio recorded) about these performances as a culmination of their learning experience.

This was a research project that was arts-based but also incorporated a focus on species and shared environments from a scientific perspective. Audio-recorded interviews, video and photographs of student artefacts (their puppets under construction and in final form) were analysed using thematic analysis, with a focus on 'examples' that gradually emerged for us as interesting and provocative instances (in terms of how we considered the student-puppet relationship) as we became increasingly familiar with the data set (Ferguson et al., 2019). Five different members of the research team undertook this data analysis, which involved individually identifying thematic units that emerged, as well as sharing them with each other and discussing and consolidating relevant issues, themes and topics.

Findings: They're Alive!

As we engaged with the extensive and rich data set, we realized that puppets were highly metaphoric in their potential for meaning making and providing opportunities for students to 'be' in the world in new ways (Gross, 2012; John & Carol, 2011), through the materiality and embodiment (discussed in the first paragraph of the 'Methodology' section) afforded by the puppets and the drama performances. Puppets are unique material assemblages that when 'activated' by puppeteers are able to enable performers to act, think and do in ways not possible when disconnected from the puppet. In this way, puppets are 'tools' like any other physical or conceptual element of our immediate environments, and as with all tools they serve a particular purpose (as we will show). Therefore, the analysis we focus on in this chapter is that which supports and reflects the embodied nature of the puppets that students created from their learning about endangered animals, as well as their embodied material engagement in creating these trash puppets and a 'theatre in a suitcase' to communicate the plight of their assigned endangered animals.

Metaphors in Action: Puppets as Representations with Epistemic Import

The research that the students conducted and the various lessons they undertook enabled them to learn about their endangered animal. In this way, as will be outlined below, the puppets as a particular type of representation, functioned to support students to develop knowledge about, and a meaningful connection with, their endangered animal; the puppets were epistemically important for the students. As the sequence of sessions unfolded, students grew closer to their animals (in puppet form) as they formed an attachment and identified with the animals in material form. This was observed at the start of the Week 3 session, during learning about skeletons and their role in puppetry with the facilitating scientist-artist. This learning was preceded by research and learning sessions at the Zoo in Weeks 1 and 2 with scientists and educators. The data analysis shows how the scientist-artist ran this session to engage the students and make the topic relevant, such as by making links to the skeletal and joint functions of endangered animals and students' own observations of familiar animals (e.g. cicada shells) or their own bodies and those of animals in the excerpt below. In doing so, the scientist-artist also used language that was common to both the animals and to human bodies such as 'skeletons' and 'joints':

Scientist-Artist: So, what sort of things do we think a skeleton does. Does anybody want to suggest? What does your skeleton do for you?

Student: Structure.

Scientist-Artist: Structure–yes, great, absolutely, anything else?

Student: Protects the organs?

> Scientist-Artist: Yes exactly, protects very important organs. Imagine if you had your brain without your skull–it would be bad. So, it's protection against dangers, against the environment…Anyone else? …(*no reply from class*). It does a number of things so support as you've suggested, protection, so it allows anchor points for muscles…Your muscles move by being anchored against bones. So, they [muscles] don't just hang out in space, they are actually joined onto bones which helps you to have a mobile body. The skeleton also grows as the body grows so as you get bigger to being an adult, your skeleton is getting bigger and bigger. Insects, although their skeletons are on the outside can also grow so if any of you have seen cicada shells lying around that shows you they have to shed their whole skeleton in order to grow to the next stage.

Both science and art education involve teaching and learning conceptual knowledge that is unique to the discipline. Wegner and Nückles (2015, p. 624) explain that conceptual metaphors emerge from particular "concepts of knowledge and knowing"– "as schemata or scripts that can be localized within an individual". The trash puppets project incorporated conceptual knowledge from art and science in the design of the program. As the above excerpt shows, students were able to translate knowledge from their research into endangered species, to ways of expressing, performing and communicating this knowledge through the arts. In doing so, conditions were set up for conceptual metaphors to be explored by students in the design and delivery of the program.

Participation metaphors emerge through "knowing as an activity (as opposed to knowledge as an entity)" (Wegner & Nückles, 2015, p. 624). Students began the project with their own knowledge and were taught some knowledge that was intended to generate conceptual metaphors for students. However, as they learned, participation metaphors emerged and as Wegner and Nückles (2015, p. 624) explain, participation metaphors are those that "cannot be separated from the context in which (they are) enacted". Indeed, participation metaphors could be seen from our analysis as materialising out of the context of students engaging in the art and drama process–making and enacting them in performances.

Explicit teaching in Week 3 was first focused on exploring a range of animal puppets (sock puppets, full-scale War Horse & insects) as a PowerPoint presentation, to demonstrate different ways of expressing emotion, likeness and movement with puppetry. Skeletal knowledge in relation to animal evolution and puppet design and function was also explained and illustrated. The next part of this Week 3 session segued into students engaging with materials as they explored, experimented and connected with recycled materials and tools strewn across two tables. This was immersive materiality, made possible by the bodily interactions of the students with the materials and their physical enactment of the technical processes-possibilities as they started to think and make their trash puppets.

The corporeal nature of 'bringing to life' these puppets was evident in the students' learning and communication of their learning, evident in the trash puppets as part of their 'theatre in a suitcase'. An outcome of the status of the puppets as material-metaphoric representations was that while needing to refer to/represent the key features of the animal itself in a scientific sense (analogical icons), they were also necessarily self-referential in an artistic sense (metaphorical icons) (Anderson, 1987). It is in this way that the transformative power of the puppets emerged from their nature as simultaneously: (1) scientific objects–in their referring to the natural world of structure, function and habitat, and (2) artistic objects–in their referring to themselves as unique material-embodied animals.

For example, metaphoric associations developed from students' connection to their animal in a 'shared world' (in this case Victoria, Australia) that extended to the social as well as shared natural environment. An example of this is how students conveyed in puppet form a 'lazy frog' that doesn't move:

> Student 3: …We used a couch (in the performance) to emphasise that it is a lazy frog 'cause in the snowy season they just lay on their backs–so it doesn't really do a lot
>
> Researcher: So, you're really emphasising its laziness?
>
> Student 3: Yeah 'cause that's one of its main traits so we just wanted to emphasise that and play on that–because it's pretty relatable.
>
> Researcher: And were you considering your audience when you…
>
> Student 4: Yeah, we made it more of a relatable frog because if we just said it laid on its back people wouldn't pay as much attention, so we made it a couch potato who likes to play footy and watch it.

Here, as students developed their understanding of the *Baw Baw Frog* (Fig. 2a) and its form and movement (or lack of), they creatively developed (along with their script) a frog in the colloquial Australian habitat of a lounge room, with a couch and television for watching football.

Empathy was evident here also–particularly when students related the demise of this frog to 'laziness' and linked this to the Australian colloquial term 'couch potato'. This indicated a translation from: (1) knowledge of the frog, its laziness and therefore its demise in its human-induced rapidly changing (for the worse) Bogong High Plains habitat; to (2) the demise of the human who has developed sedentary behaviours with the development of sedentary-inducing technology (e.g. televisions and couches), leading to the term 'couch potato'. The students do not mean 'lazy' literally, but rather use 'lazy' as a linguistic signifier; "metaphoric meaning is enhanced when language is understood through analyzing the characteristics associated with words and not necessarily just the definitions of words" (Jensen, 2006, p. 8) As is made evident by this metaphoric example, there is a shared context–a particular environmental/social place (Victoria) within Australia.

Considering Audiences: The Appeal of the Dramatic

In the process of developing the puppets for their theatre performances along with their scripts, students were considering how they would communicate with, and entertain, their target audiences (primary and pre-school children):

> Student 5: Well we were told in the brief [the presentation in Week 8] it wasn't going to be to older kids it was going to be little kids, so that was something we had to take into consideration. If this was to be pitched to older kids we would have definitely included more scientific terms and perhaps actually named the infection that is killing these frogs [Baw

> Baw frog]. But communicating science was something that, well I think it's something that can be communicated to any age group because science is just making sense of something really and by doing that with little kids we really had to think about how their minds work and what engages them and colour, and actually using a puppet can actually change the way something is portrayed.

Perhaps if they were engaging with an older audience or teachers for assessment, students would be directly communicating scientific knowledge in an abstract form. But this project required students to choreograph the dramatic play to appeal to the younger audience, and to support this audience to relate to the endangered animals in their performative puppet forms.

Students appreciated the importance of making their puppet life-sized in order to communicate their animal's story and perform (enliven) their puppets, but also how this could invoke a desire in the audience 'to learn more about it'.

> Student 1: ...for the bandicoots [Eastern Barred Bandicoot], we haven't seen them in person so we did a lot of research actually looking at photos and there's actually some videos about how they move so we could really show that through the puppeteering. So, I think making it in person is really good because this [the puppet of the bandicoot] is life size. Also, for our audience it's really useful for them to picture it in their heads.
>
> Student 2: Yeah–it's an engaging way to kind of communicate what we've learnt with the students and stuff because obviously seeing a puppet they're interested and want to learn more about it.

This suggests that students know, maybe from their own experiences, that if they are interested and engaged they are more likely to learn, so they wanted to create this learning opportunity for their audience. But more than this, interest and engagement also nurture the human-animal connection, as the students came to realise in the performance context as actor-puppeteer, puppet and audience-participant came together.

Enlivening As/With Materials

Students encountered learning as puppet construction and dramatic performance involving materials and tools. Materiality was evident (see Fig. 1) as an intimate connection between students' bodies (mainly hands) and the various materials they engaged with to make recycled materials into something meaningful (in terms of advocating for their endangered animal). They were immersed in a world of materials full of creative struggles as they needed to make decisions, take risks, experiment and communicate (often with frustration and humour) as they sought to motivate each other to progress with the task.

This development, from a lifeless collection of material components in the Week 3 session to an enlivened agent in puppet form at the Zoo's 'theatre in a suitcase performances' (Fig. 2a, b, c), was made possible by the activation of the potential life force of the materials through the bodily processes of the students in performance. During the construction process (Fig. 1), the students and the puppets are separate (although intimately connected through the students working with the materials), whereas in the performances they become one through/as performance. The material,

Fig. 1 Students starting to create trash puppets

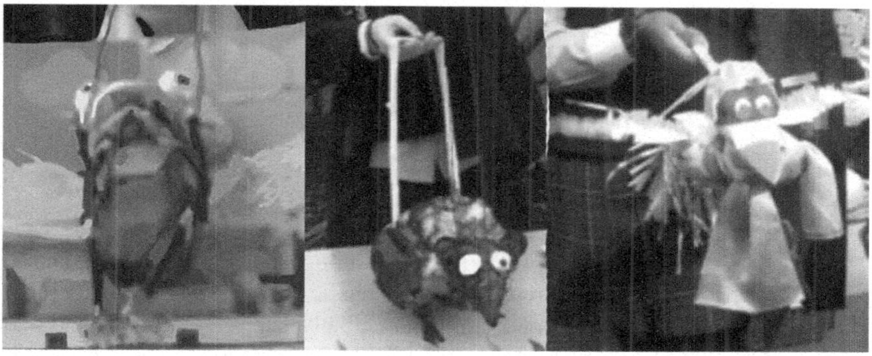

Fig. 2 Puppets 'coming to life' in the 'theatre in a suitcase' (**a** Baw Baw Frog; **b** Eastern Barred Bandicoot; **c** Helmeted Honeyeater)

performative and embodied are all present; the student in 'operating' the puppet, becomes one with the puppet through this connection of bodies. The student and puppet bodies become one anima–something different, a new soul emerges–that develops from the material connection between student and puppet in the construction process.

The *Eastern Barred Bandicoot* trash puppet (Fig. 2b) is a good example of how the puppets only came 'alive' when students and puppets connected. It was not just a matter of the student holding particular parts of the puppet in order to determine particular movements, but rather in connecting with the puppet through the strings and the handle (material), the students 'channeled' what it would be to move (bodily) as an *Eastern Barred Bandicoot*. When the students enacted these movements as/with the puppet, it–the puppet as a particular material manifestation of the animal infused with the students' understanding of this specific species–was 'telling' the students how it should move. There was thus a dialogue between student and puppet, and thus animal. The student and puppet were idle–in terms of the main aim of engaging an audience to care about endangered animals–until they connected in this way. And when they did connect, the student and puppet as bodily and material forms fused to create a new anima; the *Eastern Barred Bandicoot* as an endangered animal became 'present' to the audience.

> Adult audience member 1: Turn his face around, he's really got a, look, he really looks like he's looking at you. Awww!

The audience in this way could emotionally connect ("Awww!") with the animal/puppet as they empathise with the plight of the endangered animals. Similarly, when the students 'brought to life' the *Helmeted Honeyeater* puppet (Fig. 2c), the adults in the audience were not simply impressed with the physiological realism of the material assemblage in action, but were delighted to be present with an 'alive' bird that was its own individual and not merely a (material) representation of the abstracted features of its species:

> Adult audience member 1: Ah that's cool! So, you can actually, move its head around and make its arms flap at different rates.
>
> Adult audience member 2: Yeah that's good.
>
> Adult audience member 1: Yeah.
>
> Researcher: Yeah, the engineering is fantastic.
>
> Adult audience member 1: It's really good. It's got personality too, it's amazing how much personality they all have!

The *Helmeted Honeyeater* puppet (Fig. 2c), as with all the puppets, was a science object; an analogical icon that represents in realistic ways the natural structure–function of the endangered animal: "You can actually, move its head around and make its arms flap at different rates". But the *Helmeted Honeyeater* puppet was also experienced by the audience as 'alive' in the sense of exhibiting a distinct and individual presence: "it's amazing how much personality they all have!". This was a life force and energy that was not reducible to the animal's structure and function; this was the *Helmeted Honeyeater* puppet as art object, a self-referential metaphorical icon, it had a 'personality'.

Conclusion: Metaphor, Materiality and the Body as Performance

In this chapter, we have shared key aspects of the design of this art/science learning program and some of the making and learning experiences of those involved, particularly for the benefit of those working in the field of art/science education. We suggest that because puppets are metaphors (Anderson, 1987; Gross, 2012; Hausman, 1989; John & Carol, 2011), when students work toward a project that involves transforming and conjuring an assigned endangered species into a puppet, that embodied metaphor is at play in the creative learning process. In embodying the puppets as art objects, the students were referring to the puppets themselves and their idea of the puppet, as this was the only way they could 'bring them to life'. Puppets representing endangered species are art objects can thus be understood as metaphorical in their iconic nature. When a puppet is made, performed and/or observed by an audience, it can operate as an embodied metaphor of selves and others (human and non-human animals) in our 'shared world'.

Puppets of animals as objects are thus icons, representing not just the natural structure–function of what in this example were endangered animals (as analogical icons of science) but also referring to themselves (as metaphorical icons of art). We only discuss some of the research findings from this 8-week learning program in this chapter, but note that the coming together of the students and puppets to transcend both the student as organic creator and puppet as inorganic (material) artefact, led to the emergence of something new/different which provided opportunities for the audience to connect on an emotional level with the animals. This was not just a student 'controlling' a puppet, but the emergence/presence of something 'alive'. We suggest that the puppets as objects of science–despite their structural/functional accuracy as analogies–were not capable of engaging the audience and advocating for the represented animal. It was only as metaphors–as art objects that reference themselves in/as performance–that the puppets could realise this potential to 'come alive' for the actor-puppeteer and audience-participant. And it is in this way that important opportunities for drama and science education might be realised.

In their nature as objects of art (drama) and science, we also suggest that puppets can serve a therapeutic role as well as an educative role. We follow Bernier (2005) in considering these complimentary and mutually constitutive in young peoples' coming to terms with the confronting realities of the Anthropocene and their desire to realise new futures. In regard to the plight of endangered animals, young people can not only learn about the structure and function of these animals and what is needed to 'save' them, but also embrace their emotions in response to this potential loss and harness this emotional energy for making change. In other words, we propose that puppets as art(drama)/science objects can potentially play an important role as 'ecotherapy' (Hasbach, 2015, p. 205) in addressing the ecological grief and anxiety of young people that they experience in the face of climate change and other human-induced catastrophes, with teachers serving an important supporting role. By 'addressing' we do not mean suppressing or moving beyond, but rather embracing the transformative power of grief (Butler, 2004) and coming to terms with our 'solastalgia' which is the "pain or sickness caused by the loss or lack of solace and sense of isolation connected to the present state of one's home and territory" (Albrecht, 2005, p. 45). As we consider it, ecological grief and anxiety are not pathological responses to living in the Anthropocene, but rather a "natural response" (Cunsolo & Ellis, 2018, p. 275). This therapeutic potential of puppets as metaphors that we touch on needs much further research. As we have endeavoured to show in this chapter, it is the metaphorical nature of puppets as part of the fusion of art and science that might make it possible for students to experience their totemic-selves as they connect with the animals of our 'shared world' through performance, something that should be considered in future art (drama)/science education programs.

References

Ackerman, T. (2005). The puppet as a metaphor. In M. Bernier & J. O'Hare (Eds.), *Puppetry in education and therapy: Unlocking doors to the mind and heart* (pp. 5–12). AuthorHouse.

Albrecht, G. (2005). 'Solastalgia': A new concept in health and identity. *Philosophy, Activism, Nature, 3*, 41–55.

Amsen, E. (2019, June 14). Science educators are turning to the performing arts. *Forbes*. Retrieved from https://www.forbes.com/sites/evaamsen/2019/06/14/science-educators-are-turning-to-the-performing-arts/#4120abf737b1

Anderson, D. R. (1987). *Creativity and the philosophy of C.S. Peirce*. Kluwer Academic Publishers.

Berger, C. (2019). Introduction. In C. Berger (Ed.), *Conceptualism and materiality: Matters of art and politics* (pp. 3–14). Brill.

Bernier, M. (2005). Introduction to puppetry in therapy. In M. Bernier & J. O'Hare (Eds.), *Puppetry in education and therapy: Unlocking doors to the mind and heart* (pp. 109–116). AuthorHouse.

Bradley, J. (2020, January 11). Terror, hope, anger, kindness: The complexity of life as we face the new normal. *The Guardian: Australian Edition*. Retrieved from https://www.theguardian.com/australia-news/2020/jan/11/terror-hope-anger-kindness-the-complexity-of-life-as-we-face-the-new-normal

Butler, J. (2004). *Precarious life: The powers of mourning and violence*. Verso.

Cohen, M. (2007). Puppetry and the destruction of the object. *Performance Research, 12*(4), 123–131.

Cool, D., & Frost, S. (2010). Introducing the new materialisms. In D. Cool, S. Frost, B. Cheah, M. Orlie, & E. Grosz (Eds.), *New materialisms: Ontology, agency and politics* (pp. 1–43). Duke University Press.

Cunsolo, A., & Ellis, N. R. (2018). Ecological grief as a mental health response to climate change-related loss. *Nature Climate Change, 8*(4), 275–281.

Ellis, N., & Cunsolo, A. (2018, April 5). Hope and mourning in the Anthropocene: Understanding ecological grief. *The Conversation: Australian Edition*. Retrieved from http://theconversation.com/hope-and-mourning-in-the-anthropocene-understanding-ecological-grief-88630

Ferguson, J., Aranda, G., Tytler, R., & Gorur, R. (2019). Video research—Purposeful selection from rich data sets. In L. Xu, G. Aranda, & D. Clarke (Eds.), *Video-based research in education—Cross-disciplinary perspectives* (pp. 124–139). Routledge.

Gross, K. (2012). *Puppet: An essay on uncanny life*. University of Chicago Press.

Hasbach, P. H. (2015). Therapy in the face of climate change. *Ecopsychology, 7*(4), 205–210.

Hausman, C. R. (1989). *Metaphor and art*. Pennsylvania State University.

Head, L. (2016). *Hope and grief in the Anthropocene: Re-conceptualising human-nature relations*. Routledge.

Hickman, C. (2019, September 15). I'm a psychotherapist–Here's what I've learned from children talking about climate change. *The Conversation: Australian Edition*. Retrieved from https://theconversation.com/im-a-psychotherapist-heres-what-ive-learned-from-listening-to-children-talk-about-climate-change-123183

John and Carol (2011). *Figures of speech puppet as metaphoric being* [Streaming video]. Retrieved from https://www.youtube.com/watch?v=MZfBROpj8_w (note: the full names for the puppeteers are not available on any internet or library searches)

Jensen, G. (2006). Metaphors as a bridge to understanding educational and social contexts. *International Journal of Qualitative Methods, 5*(1), 1–19.
Kemmis, S. (2019). *A practice sensibility: An invitation to the theory of practice architectures.* Springer Nature Singapore Pte Ltd.
Latour, B., & Woolgar, S. (1986). *Laboratory life–The construction of scientific facts.* Princeton University Press.
Lemke, J. (2015). Feeling and Meaning: a unitary biosemiotics account. In Trifonas, P. (Ed.), *International handbook of semiotics.* Springer.
Mannix, L. (2020, January 8). 'Many, many billions' of animals feared to have dead in bushfires. *The Sydney Morning Herald.* Retrieved from https://www.smh.com.au/national/many-many-billions-of-animals-feared-to-have-died-in-bushfires-20200108-p53pvk.html
Matthews, J. (1998). Somatic knowing and art education. *Marilyn Zurmuehlen Working Papers in Art Education, 10*(1), 89–94.
Melbourne Zoos. (n.d). *Fighting extinction: Priority native threatened species.* Retrieved from: https://www.zoo.org.au/fighting-extinction/priority-native-threatened-species
Mello, A. (2016). Trans-embodiment. *Performance Research, 21*(5), 49–58.
Molnar, D. (2020, May 5). Climate grief and embracing beautiful confusion: Daniela Molnar interviewed. Retrieved from https://variablewest.com/2020/11/26/interview-daniela-molnar/.
O'Hare, J. (2005). Introduction to puppetry in education. In M. Bernier & J. O'Hare (Eds.), *Puppetry in education and therapy: Unlocking doors to the mind and heart* (pp. 1–4).
Oar, E., & Taylor, M. (2020, January 9). Eco-anxiety climbs as fires, smoke and animal deaths trigger fear and trauma. *The Lighthouse.* Retrieved from https://lighthouse.mq.edu.au/article/january-2020/eco-anxiety-climbs-as-fires,-smoke-and-animal-deaths-trigger-fear-and-trauma
Rosen, J. (2020, January 11). An artist set out to paint climate change. She ended up on a journey through grief. *Los Angeles Times.* Retrieved from https://www.latimes.com/la-sci-col1-climate-change-art-2019-story.html
Simon, S., Naylor, S., Keogh, B., Maloney, J., & Downing, B. (2008). Puppets promoting engagement and talk in science. *International Journal of Science Education, 30*(9), 1229–1248.
Sontag, S. (1966). One culture and the new sensibility. In *Against Interpretation* (pp. 293–304). Octagon Books.
The Climate Museum. (2020, May 5). Retrieved from climatemuseum.org.
Trommer-Beardslee, H., Dasen, A., Pangle, W., & Batzner, J. (2019). Team Hyena Puppet: An interdisciplinary approach to making and teaching science through art. *Teaching Artist Journal, 17*(1–2), 45–50.
Vince, G. (2020, January 13). How scientists are coping with 'ecological grief'. *The Guardian: Australian Edition.* Retrieved from https://www.theguardian.com/science/2020/jan/12/how-scientists-are-coping-with-environmental-grief
Wegner, E. & Nuckles, M. (2015). Knowledge acquisition or participation in communities of practice? Academics' metaphors of teaching and learning at the university. *Studies in Higher Education, 40*(4), 624–643.

Shelley Hannigan (BFA, M.Ed, GradDipEd, Ph.D) has been an academic at Deakin University specialising in visual arts, arts education and developing new units in creative education since 2005. As an artist her focus has been on experiential and experimental processes of creating art with a focus on identity, place and ecology. Her research draws on her practice work from multiple fields, with a particular interest in interdisciplinary and transdisciplinary phenomena. Her PhD research investigated the phenomenon of place and identity in artistic practice and thinking. Her research in art-science education, art and wellbeing, incorporate this research of art, place and identity to highlight the value and usefulness of art in education and other practices.

Joseph Ferguson (BA, B.Sc, M.Teach, Ph.D) is an educational researcher and teacher educator interested in investigating the various manifestations of reasoning in the science classroom, in particular creative reasoning. Through an exploration of the philosophies and theories that underpin inquiry practices and the associated methodologies that enable this research, Joseph seeks to add to efforts to clarify what it means for teachers and students to enact inquiry in the classroom. Joseph is particularly interested in the power of video-based methodologies, informed by film theory and philosophy, to maximise the potential of such research to support teachers and students in their teaching and learning of science.

"This is the Funniest Lesson": The Production of Positive Emotions During Role-Play in the Middle Years Science Classroom

Senka Henderson and Donna King

Abstract This study builds on our previous work where specific science activities, such as demonstrations and laboratory activities, evoked students' positive emotional responses and focused students' attention on the science content they were learning. In this study we were interested in students' discrete emotions in a Year 8 science class expressed during a role-play activity in a biology unit on skin burns, called 'Singed'. Data from two focus groups of students from the class are presented. Drawing on multiple data sources, including classroom video recordings, observations of classroom transactions, thinking prompts, field notes and emotion diaries completed at the end of each lesson, we developed insights into individual student emotions. Using a theoretical perspective drawn from theories of emotions founded in sociology, we identified that students expressed the emotions of happiness, joy, and enthusiasm during the role-play. These positive experiences aligned with a high interest score reported by students when the class results were averaged. Importantly, the thinking prompts which were questionnaires completed before and after the role-play, showed evidence of students' learning and understanding of the science concepts related to skin burns. This study suggests that role-play can be used successfully as a teaching strategy in the middle years.

Keywords Role-play · Discrete emotions · Interest · Middle school science · Emotion diary

S. Henderson (✉)
Faculty of Education, Queensland University of Technology, Brisbane, Australia
e-mail: s10.henderson@qut.edu.au

D. King
School of Education, Faculty of Education & Arts, Australian Catholic University, Brisbane, Australia
e-mail: Donna.King@acu.edu.au

P. J White et al. (eds.), *Science and Drama: Contemporary and Creative Approaches to Teaching and Learning*, https://doi.org/10.1007/978-3-030-84401-1_11

Introduction

There is a growing concern internationally and in Australia about the disengagement and decline of interest and motivation of science students in the middle years (i.e. from age 11 onwards) (Anderhag et al., 2016; Jenkins & Nelson, 2005; Osborne & Collins, 2001; Potvin & Hasni, 2014; Potvin et al. 2020). This is an important issue for science educators, as disengaged middle years students will be less likely to enrol in senior science subjects (Kennedy et al., 2014; Osborne et al., 2003) and consequently, less likely to develop informed science perspectives, become scientists, or science-trained professionals (Tytler, 2007). Of further concern, is the PISA (Program for International Student Assessment) 2018 survey indicating that the performance of Australian students in mathematics and science has been steadily declining between 2003 and 2012 (Thomson et al., 2019). Furthermore, according to Sue Thomson, the PISA national project manager for Australia, this finding is of "a concern … (as) we're not giving them [students] the same level of skills as they are in other countries and [it is] a 'wake-up call'" (Duffy & Wylie, 2019, para. 5). The PISA study also indicated that when compared to the average student across OECD counties, Australian students felt more afraid of failing science (Thomson et al., 2019).

To increase engagement and enjoyment in science in the middle years, there is a need to implement effective pedagogical approaches and simultaneously to understand the role of emotions expressed by students during those approaches. Sinatra et al. (2014) noted that "in order to broaden participation in science, we must capitalise on student emotions that are adaptive for science learning and those that promote sustained interest and pursuit of science careers" (p. 415). Students can be engaged in emotionally engaging, drama-based pedagogies (i.e. role-play), that are fun and stimulating, but further research is needed to understand better if these lessons contribute to a student's positive perception of science.

Since the relationships between drama-based pedagogies (DBP), science and student emotions are not fully understood, this study will examine the interplay between the 'organ donation' role-play and students' discrete emotions and self-reported interest. One broad research question that guided the research design was: Does the pedagogical strategy of role-play make a difference to students' emotional experience of science? As the study progressed, two more focused questions emerged:

1. What were the positive emotions experienced by the students during the role-play activity?
2. What evidence is there of students' understanding of science terms/concepts?

In order to understand better the interplay between role-play and emotions, in the sections that follow, we explore students' emotions expressed during the science activity and we provide theoretical perspectives on emotions research.

Emotions Research

Previously, science education research has mostly focused on the effect of logical reasoning and cognition in learning science. However, over the past 15–20 years, research on emotions in education has emerged and highlighted the importance of understanding emotions experienced by students, and the significance of emotions on learning outcomes (Ritchie & Tobin, 2018; Sinatra et al., 2014). Immordino-Yang and Demasio (2007) have noted that "when we educators fail to appreciate the importance of students' emotions, we fail to appreciate a critical force in students' learning" (p. 9).

Our previous research has shown that middle school students experience both positive and negative emotions when learning science (King et al., 2015, 2017). Research on positive emotions has contributed considerably to understanding the nuances of positive emotional arousal in science classrooms. For example, positive activating emotions, such as enjoyment are associated with positive outcomes (Pekrun, 2006). Research has shown that positive affect increases engagement because positive emotions, such as happiness and joy help to "broaden one's thought action repertoire and build resources" (Fredrickson, 2001, p. 221). One study by Tomas and Ritchie (2012) has shown that students experienced pride, strength, determination, interest and alertness during the BioStory project (a science project with a moral and ethical dilemma) and those emotions were associated with enhanced feelings of self-efficacy. Bellocchi and Ritchie (2015) found links between pride and triumph within classroom interactions and instructional tasks during learning episodes on the topic of energy. This study builds on our earlier research (King et al., 2015) by investigating if a role-play activity has an effect on middle years students' discrete emotions and interest when acting out a scenario related to the socio-scientific issue of organ donation.

Pedagogical Approaches Research: Why Use Role-Play in Science?

Other research has focussed on novel pedagogical approaches used in science classrooms to engage students and generate interest in the problem presented, for example, using video games such as *MinecraftEdu* (Pusey & Pusey, 2015), using gamification (Fleischmann & Ariel, 2016), or using historical vignettes (Bellocchi, 2004) in the science classroom. This research has shown that through novelty and fun, those approaches can generate interest, enhance teaching and learning of science concepts and help students to develop a deeper understanding of the related science concepts and increase their motivation and engagement in science.

Research has also shown that using dramatic pedagogies, such as role-play can be utilised in classrooms as a different pedagogical approach (Dorion, 2009; McSharry & Jones, 2000; Ødegaard, 2003). Role-play is a drama-based pedagogy (DBP) focused on an embedded process-oriented approach to learning (Lee et al., 2015). The potential benefits of role-play for learning and teaching have long been recognised. It is a fun, engaging and stimulating activity that can motivate students to learn. During role-play, the learners who are participating are required to immerse themselves in a social scenario and take on a role in which they pretend to be someone or something other than themselves (Killen, 2013). Through role-play students are developing their creativity and imagination, which assists in the development of scientific thinking and conceptual understanding and makes learning in science more attractive, especially to students who have been disengaged with science (Heyward, 2010; Killen, 2013). Role-play can be used in teaching of emotionally or behaviourally difficult to handle real-life topics, such as sex and drugs education, child-protection (McSharry & Jones, 2000), or to teach students about moral or ethical issues arising from the curriculum (Colby, 1987), such as climate change, nuclear technology, or organ transplantation. However, there is little research on role-play in school classrooms and the emotions expressed by students as they enact scenarios related to socio-scientific issues. Serious science topics can use role-play to elicit a variety of emotions in students. Positive emotions expressed during the role-play of serious topics may enable students to participate more fully in learning rather than withdraw due to the confronting nature of the topic.

Theoretical Perspectives

This study was informed by theoretical perspectives drawn from theories of emotions founded in sociology (Turner, 2007, 2009). Turner's (2009) sociological theory of emotions is founded on the premise that the dynamics of specific emotions and the social organisation that causes the arousal of discrete emotions are important in theorising about human emotions. He explains that emotions are produced in "sociocultural conditions and once aroused [will] have effects on these conditions" (p. 342). In a review of twenty scholarly readings, Turner suggests that there are four primary emotions: anger, fear, sadness, and happiness. Emotions are also valanced, i.e. they can be categorised as positive and negative (Stets, 2010). In humans, emotions can be aroused in varying levels of intensity from low, medium through to high-intensity states. Happiness/joy and enthusiasm would be valanced positive and embarrassment would be negative (Turner, 2007). Turner argues that the sociocultural environment impacts on students' expression of emotions, which affects the dynamics of face-to-face encounters and the larger social structures in which they occur. Turner asserts that understanding the sociocultural origin of discrete emotions may afford knowledge about how these emotions affect the micro- (e.g. in face-to-face interactions), meso- (e.g. classroom procedures), and macro- (e.g. school policy) levels of social reality.

Methods for Identifying Emotions in Grade 8 Science

This study is informed by Stake's (2005) interpretive perspective of case studies in conjunction with understandings of ethnography (Creswell, 2012). We have attended 26 lessons over a 9-week period. The lessons were approximately 50 min long. There were 27 students in the class. The unit called 'Singed' required students to learn about skin transplants after they complete a story where a boy is burnt when a jet ski explodes. Drawing on the analysis of multiple data sources, including classroom video recordings, observations of the classroom, field notes, emotion diaries completed at the end of each lesson, thinking prompts and the analysis of facial expressions, insights into individual student's emotions are reported. The analytical techniques used are described in more detail below (i.e. meso-level and micro-level analysis).

Meso-Level Analyses: Emotion Diary (EmoDiary)

Based on the work of Zembylas (2008) and Ritchie (Ritchie et al., 2016) we included 10 discrete emotions on the emotion diary (EmoDiary). Those emotions were agreed by a panel of researchers as the most salient emotions for the middle years' students. They included happiness/joy, sadness/disappointment, anger/irritation, anxiety, disgust, pride, wonder, enthusiasm, frustration, and embarrassment. To prevent students' confusion of less familiar labels, an emoticon for each label was added. At the bottom of the EmoDiary was a scale where students rated their interest level in each lesson on a scale from 1 (very bored) to 10 (very interested). A mean score of interest for each lesson was calculated by averaging students' responses on the interest scale.

Students were asked to complete an EmoDiary at the end of each lesson, indicating any emotion that was strong enough for them to notice and to indicate the intensity of the particular emotion (low, medium, high). During the period of the data collection we accessed over 500 8th grade students' diaries.

An EmoDiary is a self-reporting instrument that requires student cooperation for completion. We met the students during the first lesson at the beginning of the 'Singed' unit to introduce the research and explain the EmoDiary and how students should identify their emotions. The students were given a practice EmoDiary, and following a whole class discussion we were confident that students understood how to recognise emotions and to fill in the diary. The EmoDiaries were analysed by counting each of the emotions reported by students, lesson by lesson. These were graphed and trends were found which can be seen in Appendix 1. Students' comments on the EmoDiaries were analysed thematically.

Meso-level Analyses: Thinking Prompts

The Thinking Prompts were a paper-based questionnaire that examined students' progressive understanding of science concepts which were administered to students on four different occasions in the following lessons: week 1/lesson 4; week 3/lesson 4; week 4/lesson 4; week 8/lesson 3. This enabled us to examine changes in students' answers over the duration of the unit.

In the Thinking Prompts, students were asked eight questions which related to unit content. For example:

1. Have you heard of first, second and third degree burns before?
2. What evidence is used to determine the type of skin burn?
3. What are the differences between them?
4. Have you heard of organ harvesting?

Student responses to the questionnaire were analysed qualitatively in order to gauge students' conceptual understanding that could be attributed to their participation in various lessons using a wide variety of pedagogical approaches, including role-play throughout the duration of the unit.

Micro-level Analyses: Facial Expression

Analysis of the video recordings have been used to complement our data drawn from the EmoDiaries and Thinking Prompts. The videos were analysed for expression of students' emotions. The manual procedure for interpreting students' facial expressions developed by Ekman and Friesen's (2003) Facial Action Coding System (FACS) have been applied. This analysis of in-the-moment students' emotional facial expressions was used to reinforce or triangulate the discrete emotions identified from the multiple methods utilised in this study.

Study Context: The School, The Biology Unit, The Two Case Studies

This study was conducted in a Year 8 science classroom in a large co-educational urban school in South East Queensland. The science class consisted of 13 boys and 14 girls, typically aged between 12 and 13 years. The context for this science unit was burns. The unit explored concepts related to burns such as the structure of cells and function, skin structure, skin grafts, organ harvesting and organ donations affording opportunities for connections with the real-world. Lessons included a variety of activities such as computer-based lessons where students researched organ harvesting, teacher-led lessons, lessons with a guest speaker, debates, video lessons, decision making activities, poster activities, reporting, role-play activities and laboratory activities. During the unit, an innovative story-writing approach about socio-scientific issues, namely BioStories was used (e.g. Ritchie et al., 2011).

In this study, the researchers were interested in the emotional engagement of the year 8 students during the role-play activity that occurred in the unit during week 6, lesson 2. During the role-play activity the students were asked to form groups (4–6 students) and to firstly, write a script and secondly, act out a short drama depicting the ethical dilemma about organ donations through role-play. The task required students to reproduce the events that could occur when a decision was needed to be made about the use of organs from a brain damaged relative who was in hospital. Students were given four defined role-play characters that included:

1. The intensive care doctor;
2. A parent of the young person who is brain damaged from, either (a) the collapse of a rugby scrum or (b) a fall off a horse whilst show jumping;
3. A brother or sister;
4. The hospital transplant specialist coordinator.

The research presented here focuses on two focus groups where students chose their own groups for this activity. The first group consisted of high achieving students who worked very hard to achieve good results in science. Table 1 shows the names of those students and the character roles that students were acting during the role-play.

The second group were academically lower achievers than Focus Group 1, but demonstrated high levels of engagement with the activity. Names of those students and their character roles are presented in Table 2.

We decided to focus on these two groups, as they present the range of emotional experiences expressed during the activity. In addition to the fine-grained analysis

Table 1 Focus group 1

Student's name (Pseudonym)	Role-play character
Rosie	Intensive care doctor
Cassie	Hospital transplant specialist coordinator
Tessa	Sister
Alicia	Mum

Table 2 Focus group 2

Student's name (Pseudonym)	Role-play character
Alex	Intensive care doctor
Nola	Hospital transplant specialist coordinator
Carol & Vanessa	Twin sisters
Cathy	Mum
Sophie	Brain damaged person

of the two case studies, we graphed the results of discrete emotions as recorded by each student in the class across 10 of the 26 lessons. We chose to focus on those 10 lessons only, because those lessons were at the end of the unit and they were representative of the range of the emotions and interest recorded by the students. Furthermore, we graphed the average interest scores as recorded by each student for each lesson on a Likert scale at the end of the EmoDiary. The peaks and troughs of the average interest score for the entire class are in Appendix 1. Whilst a full analysis of this graph is beyond the scope of this chapter, the analysis provided an indicative view of the variation in emotions and interest ratings across 10 (out of 26) lessons of observations. Lesson 2 in week 6 stood out in the analysis of the interest graph for showing the highest average interest recorded by students (8.67/10). This high average interest score, coupled with the many positive emotions recorded by students correlated with a lesson on role-play and required further investigation to understand better students' emotions and the activities that may have led to a positive experience. In the following section we present two cases to show how the emotions were aroused in the class during the role-play activity. We also argue that through the Thinking Prompts students have shown an improved understanding of the science of skin burns developed throughout the whole unit.

Results

Assertion 1. Positive emotions of happiness/joy and enthusiasm were expressed during the planning and re-enactment stages of the role-play activities.

A summary of students' mean interest levels across the 10 lessons in the unit were calculated by averaging students' responses to the interest scale (see Appendix 1). Overall, students' mean interest scores varied between 6.88 (lesson 6.1) and 8.67 (lesson 6.2). The mean score for the entire unit (26 lessons) was 7.30, suggesting that the students were generally interested during the science lessons. The 'organ donation' role-play activity scored the highest average rating of 8.67, compared with lessons such as lecture-style lessons with PowerPoint slides (6.88) and the computer room research task (7.09). In the role-play lesson students in both focus groups expressed a high level of interest and also reported experiencing the discrete emotions of happiness/joy and enthusiasm. The following analysis shows how the students from both focus groups were expressing heightened emotions during the planning stage of the role-play.

Focus Group 1 (FG1)

This focus group consisted of 4 students, namely Rosie, Cassie, Tessa, and Alicia (see Table 1). The lesson started with the students inside the classroom. Five minutes into this lesson, students were told to go outside and practice their role-play. The following excerpt occurred 14 min and 36 s during the practice role-play session outside the classroom (see Excerpt 1). Both Cassie and Tessa were seated on the table, while Alicia was taking notes and writing the script. They were discussing organ transplant as they wrote the script and discussed death and what happens to the body. In turn 1 Alicia was asking Cassie (who was playing the hospital transplant specialist role) what she should say in the role-play when thinking about donating her organs when she dies. "What do you want done with your body when you die?" In turn 2 Cassie agrees that this is the correct way to state the question in the script rather than "what do they do with your body when you die?" This toing and froing to determine the correct script was typical of the way the students engaged with each other. In turn 3 while leaning on the table and covering her hand with her left arm, Alicia (who is playing the role of mum) says "And at that part I'll bury my head into my eyes" and started to laugh. Although the script they were creating was relatively sad (referring to Alicia's death and the removal of her body), there was much laughter. Alicia's laugher appeared contagious as other girls started to laugh in synchrony with her, showing a shared positive mood and collective effervescence (Fig. 1) (Tobin et al., 2013). However, afterwards, as Tessa sought to control her laughter, she commented that "This is awful" (turn 4) referring to the more sober topic of a discussion of a body after death. In the following turn Cassie said in a quiet voice that "This is a very interesting science lesson" (turn 5), and Rosie agreed. The excerpt below revealed a moment where Cassie states that she is finding this lesson interesting. As Cassie was a high achieving, conscientious and quiet student, her statement was noteworthy for us.

Using Turner's (2009) sociological lens that emphasises the interplay between the dynamics of specific emotions and the sociocultural conditions during the group

Excerpt 1 Focus group one planning their role-play: an 'Interesting lesson'

Turn	Speaker	Role	Conversation
1	Alicia	Mum	Yeah. What do they do with your body when you die? No, that's what you would say; What do you want done with your body when you die?
2	Cassie	Transplant specialist	Yes
3	Alicia	Mum	And at that part I'll bury my head into my eyes *(covering her head with the left arm)* into your eyes *(all laughing)* (Fig. 1)
4	Tessa	Sister	This is awful
5	Cassie	Transplant specialist	This is a very interesting science lesson
6	Rosie	Intensive care doctor	Good. Aj?
7	Alicia	Mum	Don't let me write anymore
8	Tessa	Sister	Are we done?
9	Rosie	Intensive care doctor	Is that done?
10	Alicia	Mum	No, and then you should be like, mummy I think we should do this *(all laughing)*. Yeh, I like that

Fig. 1 Girls are happy

work that contributed to the expression of these emotions, we have identified students' positive emotions during the role-play activity. Through the analysis of the EmoDiary completed by the FG1 students, it was evident that FG1 students experienced high intensity emotions of happiness/joy and enthusiasm as reported in the diary (see Table 3) during this event outlined in Excerpt 1. For example, Cassie commented that she was "Very happy today that I couldn't stop laughing while presenting the play. It was so funny that it was hard to keep a straight face", and that she was "very enthusiastic throughout [the] lesson to participate in [the] group activity and perform it, even [if] we couldn't stop laughing".

Focus Group 2 (FG2)

There were six students in the Focus Group 2 (FG2) (see Table 2). Similarly, as for FG1, the FG2 students expressed excitement and enthusiasm when they were planning the role-play. In a 4-min segment of the lesson, the teacher had to tell them four times to be quiet because their voices and laughter were becoming too loud. Eventually, the teacher asked them to move to another location where their voices would not travel to other classes. On one occasion the teacher said: "I love you being excited, but you've got to respect other people". After moving to the other location, behind the classroom, FG2 students continued with their role-play practice. The girls agreed that Sophie will play the brain damaged person after the fall off a horse. The following conversation (Excerpt 2) occurred 10 min and 36 s into their practice session when students were discussing if they should revive a dead person played by Sophie.

Despite Sophie expressing that she did not like the idea of the two boys performing CPR on her (turns 2–10), she showed happiness through her exclamation in turn 13 with the statement that "This is the funniest lesson ever" (turn 13 & Fig. 2). At this moment her facial expressions showed the emotion of happiness, as defined by Ekman and Friesen (2003), as her cheeks were raised, the corners of her lips were drawn back with mouth parted and teeth exposed. In the following turn, Vanessa (sister) agreed with Sophie and said, "It is!". Both students also indicated in their EmoDiary,

Excerpt 2 Sophie is happy

Turn	Speaker	Role	Conversation
1	Nola	Transplant specialist	Yeah but they will never recover but some other people might die because you did not donate their organs
2	Alex	Intensive care doctor	Unless you guys want to try CPR
3	Sophie	Brain damaged person	Don't you dare!
4	Vanessa	Sister	Yes?
5	Sophie	Brain damaged person	Ew. No. Gross
6	Nola	Transplant specialist	Maybe Sean?
7	Carol	Sister	Ew. Even more gross
8	Vanessa	Sister	Or Tim Fanshawey?
9	Nola	Transplant specialist	Fanshawey?
10	Vanessa	Sister	Fanshawe!
11	Nola	Transplant specialist	Okay. Then after you say yes. After you sign the papers and say yes
12	Cathy	Mum	I'm not signing anything
13	Sophie	Brain damaged person	This is the funniest lesson ever (Fig. 2)
14	Vanessa	Sister	It is!

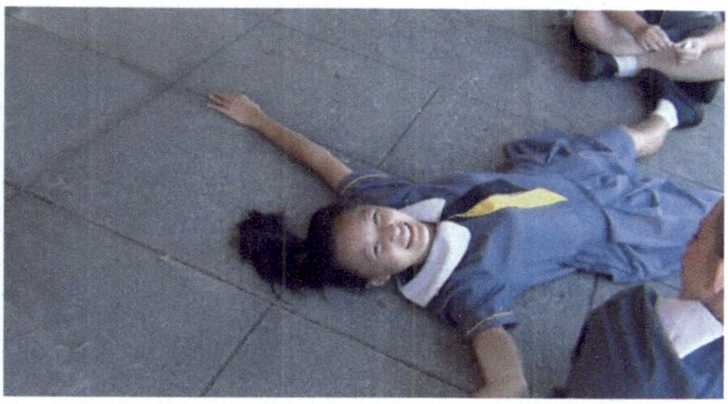

Fig. 2 Sophie is happy: "This is the funniest lesson ever"

a high level of happiness/joy and enthusiasm and commented that "Today was fun, funny". Nola who was playing the character of a specialist doctor experienced a high level of pride as reported on her EmoDiary and commented that it was the "Best play ever by us" (Table 3).

From the field notes made by the researcher (author 1) it also was evident that FG 2 students and especially, Vanessa, Sophie and Nola have experienced positive emotions, as they were laughing a lot, smiling and being emotionally involved in their role-play practice session. Further evidence of students' happiness in FG1 and FG2 can be seen in the recorded emotions and comments written on their EmoDiaries that day (Table 3). The analysis shows that students in both focus groups expressed the positive emotion of happiness/joy and enthusiasm in the role-play lesson and that

Table 3 Reported emotions by students from both Focus Groups (FG1 and FG2)

Student	Discrete emotion	Intensity of emotion (Low/Med/High)	Comments on EmoDiary to question: *In your own words, what you were doing, and what happened, when you experienced the emotion?*
Focus group 1			
Cassie	Happiness/Joy	High	Very happy today that I couldn't stop laughing while presenting the play. It was so funny that it was hard to keep a straight face
	Enthusiasm	High	Very enthusiastic throughout lesson to participate in group activity and perform it, even we couldn't stop laughing
Alicia	Happiness/Joy	High	Drama activity *(with drawing of a heart)*
	Enthusiasm	High	To drama-up science
Tessa	Happiness/Joy	High	Play/Acting
	Enthusiasm	High	Drama/skit
Rosie	Happiness/Joy	High	Role playing
Focus group 2			
Sophie	Happiness/Joy	High	Today was fun, funny
	Enthusiasm	High	
Alex	Happiness/Joy	High	REALLY FUN hahaha
	Enthusiasm	High	Coz it was fun doing the play thing
Nola	Happiness/Joy	High	Role play
	Pride	High	Best play ever by us
Cathy	Happiness/Joy	High	We did plays today!
Vanessa	Happiness/Joy	High	Role-play
	Enthusiasm	High	Role-play!
Carol	Happiness/Joy	High	Role-Play!

they rated the lesson as highly interesting. An overall interest level for both groups was 9.6/10.

In contrast, despite having a safe and supportive environment provided by the teacher and other students, three students from both focus groups experienced different intensity levels of embarrassment during the re-enactment and performance in front of the class. Alicia and Tessa were from the high achieving FG1 and both experienced embarrassment, indicating on their EmoDiaries that the intensity of it was high and commented that they were embarrassed because of "the awkward performance" and "when presenting". Nola, a student from FG2 indicated that she experienced a low level of embarrassment, with a comment in her emotion diary of

"Role play" next to this emoticon. It is important to note that only one more student from the entire cohort of 27 students indicated experiencing a low level of embarrassment. Overall, 4 students from the whole class of 27 students felt embarrassed when acting out in front of the class reminding us that uncomfortable feelings can be experienced when students are asked to perform in front of their peers. While role-play is a strategy that was experienced predominantly positively by most students in the class, we suggest that teachers incorporate strategies for encouraging students to participate who may be hesitant or embarrassed.

Assertion 2. Students showed an understanding of science concepts related to skin burns.

The Thinking Prompts, a paper-based questionnaire that examined students' progression of learning science concepts (e.g. skin anatomy and physiology, types of burns, skin grafts, organ harvesting) were administered to students on four different occasions, as explained earlier. Table 4 describes the progressive improvement of two students' understanding related to the topic of burns. Two students' responses (Cassie and Sophie) before and after the role-play are included.

Analysis of the Thinking Prompts revealed evidence of students' improved understanding of the biology concepts related to skin burns after the role-play (e.g. in

Table 4 Representative responses of the thinking prompts by Cassie and Sophie

Thinking prompts	
Before role-play (W3/L4)	After role-play (W8/L3)
(a) Cassie (from Focus Group 1)	
Q1. Have you heard of first, second and third degree burns before? What are the differences between them?	
Not sure, know that first, second and third degree comes in a particular order	First degree burns involve the injury to the **epidermis*** layer
	Second degree burns involve the **dermis** layer (part of)
	Third degree burns involve the whole entire epidermis layer and can extend through to the **subcutaneous** layer if severe
Q2. What evidence is used to determine the type of skin burn?	
(No entry)	Blisters, severe pain + red skin for 2nd degree (also swelling)
	Redness and local pain is felt in 1st degree
	No pain felt in injury + Depth of injury for 3rd degree (loss of pain sensitivity)
Q3. Have you heard of skin grafts before? What do you think it is?	
No. I don't know, maybe something to do with skin issues?	Yes. They are **donated** skin that is used on patients if their skin can't produce skin for the burn/wound to heal
Q4. Have you heard of organ harvesting? What is it?	
No. Harvesting the organs! (I think)	Yes. Practice of removing usable organs from somebody dead in order to be **transplanted** into somebody else

(continued)

Table 4 (continued)

Thinking prompts	
Before role-play (W3/L4)	After role-play (W8/L3)
(b) Sophie (from Focus Group 2)	
Q1. Have you heard of first, second and third degree burns before? What are the differences between them?	
Yes, a first degree burn is a minor. A second degree kills the skin but should heal. Third degree burns kill the nerve endings and leave scars	Well, first degree only burns the **epidermis,** the second degree burns through the epidermis and into the **dermis** and a third degree burn, burns through all the layers, including the **subcutaneous***
Q2. What evidence is used to determine the type of skin burn?	
I don't know???	How deep the burn is, the percentage of skin that has been burnt and the amount of **scaring**
Q3. Have you heard of skin grafts before? What do you think it is?	
Yeah. I don't know, I have just heard of it	Yeah! *(with drawing of a smiley face).* When skin is taken from the **donor site** & placed on the **recipient site**
Q4. Have you heard of organ harvesting? What is it?	
Yes. When someone dies, if they have good organs, they can be given to people who need them	Yeah! *(with drawing of a smiley face).* **Harvesting** deceased people's organs and giving them to people in need

Note* Words marked in **bold represent new biology science concepts

question one, both students have used a new biology term accurately for the skin layers, such as dermis, epidermis and subcutaneous layers). Interestingly, for all of the students in the two focus groups, the analysis showed that 8 out of the 10 students showed an improved understanding of the concepts of skin burns through the Thinking Prompts responses, in a similar way to Sophie and Cassie. However, we cannot say definitively that the role-play contributed to students' improved responses on the Thinking Prompts, but that role-play was one strategy used to encourage an understanding of the science concepts associated with organ transplant. Two other students were absent from the class when the Thinking Prompts survey was administered after the role-play.

Summary

In this multiple-case study we report on the interplay between role-play and student emotions. The findings in this study highlight the importance of group interactions and cooperative design during the preparation of the role-play to enable students to 'act out' the ethical dilemma of organ donation. The study has found that role-play evoked students' strong positive emotions (high intensity expressions of happiness/joy and enthusiasm) when students enacted character roles during the role-play. These positive experiences aligned with a high interest score when class results were averaged.

Importantly, the Thinking Prompts which were written before and after the role-play, showed evidence of improved students' learning and understanding of the science concepts related to skin burns (Table 4). However, it should be noted that apart from using the role-play during the unit, the teacher utilised various other pedagogical strategies, such as debates, hands-on activities, poster activities and decision-making activities.

As the research has shown, in the science classroom, the role-play is still underused and not a regular part of teachers' repertoire, often because role-play is used more frequently in creative arts-based lessons and there is a lack of examples and research showing how it can be used in the science classroom. Some teachers also fear that they are losing control in the classroom when teaching in an active way. However, this chapter indicates that the meaningful incorporation of drama-based pedagogy (i.e. role-play) with science in a year 8 class increased students' positive discrete emotions and interest while contributing to students' learning of science concepts.

We argue that role-play, as an emotionally engaging activity, should be used in the classroom to complement other pedagogical approaches in order to enhance students' engagement and interest, and that it can potentially contribute to a student's positive association with science. To make effective use of role-play in science education teachers need to be cognisant of how the role-play will support students' learning and development of conceptual science understanding and which strategies to use for encouraging students to participate who may be embarrassed. If done correctly, role-play can be enjoyable and rewarding emotional classroom experiences for the students and teachers and offers significant potential to contribute to a student's positive association with science. Such positive experiences may lead to students recalling science favourably which may have a longer lasting impact; influencing students to choose to continue studying science beyond post-compulsory schooling or for a continued personal interest in science. However, we do not advocate that all science lessons should be a highly intense emotional experiences for students, but such lessons occasionally, can contribute to positive outcomes for students.

Through a better understanding of the type of science activities that evoke positive emotions in students, and how this contributes to students' interest and enjoyment of science, we can determine the activities that are more likely to emotionally engage middle school students. Such research will be of value to current science teachers and pre-service science teachers for planning activities that afford opportunities for positive experiences for students.

Appendix 1: Graph of Mean Interest Rating and Emotions as Recorded by Students for Weeks 6–8

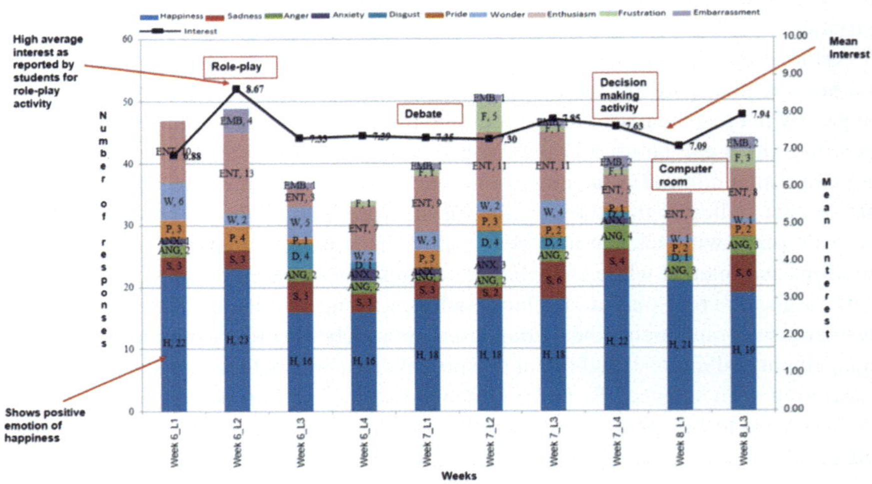

References

Anderhag, P., Wickman, P.-O., Bergqvist, K., Jakobson, B., Hamza, K. M., & Säljö, R. (2016). Why do secondary school students lose their interest in science? Or does it never emerge? A possible and overlooked explanation. *Science Education, 100*, 791–813. https://doi.org/10.1002/sce.21231.

Bellocchi, A. (2004). Designing and using historical vignettes in science teaching: A personal account. *Teaching Science, 50*(2), 14–17.

Bellocchi, A., & Ritchie, S. M. (2015). "I was proud of myself that I didn't give up and I did it": Experiences of pride and triumph in learning science. *Science Education, 99*, 638–668.

Colby, R. (1987). Moral education through drama: A 'beyond justice' perspective. *Two D Drama/dance, 7*(1), 72–80.

Creswell, J. W. (2012). *Educational research: Planning, conducting and evaluating quantitative and qualitative research* (4th ed.). Pearson Education Inc.

Dorion, K. R. (2009). Science through drama: A multiple case exploration of the characteristics of drama activities used in secondary science lessons. *International Journal of Science Education, 31*(16), 2247–2270.

Duffy, C., & Wylie, B. (2019). *Australian students behind in maths, reading and science, PISA education study shows.* Retrieved from https://www.abc.net.au/news/2019-12-03/australia-education-results-maths-reading-science-getting-worse/11760880.

Ekman, P., & Friesen, W. V. (2003). *Unmasking the face. A guide to recognizing emotions from facial clues.* Prentice-Hall.

Fleischmann, K., & Ariel, E. (2016). Gamification in science education: Gamifying learning of microscopic processes in the laboratory. *Contemporary Educational Technology, 7*(2), 138–159.

Fredrickson, B. (2001). The role of positive emotions in positive psychology. *American Psychologist, 56*(3), 218–226.
Heyward, P. (2010). Emotional engagement through drama: Strategies to assist learning through role-play. *International Journal of Teaching and Learning in Higher Education, 22*(2), 197–203.
Immordino-Yang, M., & Damasio, A. (2007). We feel, therefore we learn. The relevance of affective and social neuroscience to education. *Mind, Brain and Education, 1*(1), 3–10.
Jenkins, E., & Nelson, N. W. (2005). Important but not for me: Students' attitudes towards secondary school science in England. *Research in Science & Technological Education, 23*(1), 41–57.
Kennedy, J., Lyons, T., & Quinn, F. (2014). The continuing decline of science and mathematics enrolments in Australian high schools. *Teaching Science, 60*(2), 34–46.
Killen, R. (2013). *Effective teaching strategies: Lessons from research and practice* (6th ed.). Cengage Learning.
King, D., Ritchie, S., Sandhu, M., & Henderson, S. (2015). Emotionally intense science activities. *International Journal of Science Education, 37*(12), 1886–1914. https://doi.org/10.1080/09500693.2015.1055850.
King, D., Ritchie, S. M., Sandhu, M., Henderson, S., & Boland, B. (2017). Temporality of emotion: Antecedent and successive variants of frustration when learning chemistry. *Science Education, 101*(4), 639–672.
Lee, B., Patall, E., Caution, S., & Steingut, R. (2015). The effect of drama-based pedagogy on preK-16 outcomes: A meta-analysis of research from 1985–2012. *Review of Educational Research, 85*(1), 3–49.
McSharry, G., & Jones, S. (2000). Role-play in science teaching and learning. *School Science Review, 82*(298), 73–82.
Ødegaard, M. (2003). Dramatic science. A critical review of drama in science education. *Studies in Science Education, 39*(1), 75–101. https://doi.org/10.1080/03057260308560196.
Osborne, J., & Collins, S. (2001). Pupils' views of the role and value of the science curriculum. *International Journal of Science Education, 23*, 441–467.
Osborne, J., Simon, S., & Collins, S. (2003). Attitudes towards science: A review of the literature and its implications. *International Journal of Science Education, 25*(9), 1049–1079.
Pekrun, R. (2006). The control-value theory of achievement emotions: Assumptions, corollaries, and implications for educational research and practice. *Educational Psychology Review, 18*, 315–341.
Potvin, P., Hasni, A., Sy, O., & Riopel, M. (2020). Two crucial years of science and technology schooling: A longitudinal study of the major influences on and interactions between self-concept, interest, and the intention to pursue S&T. *Research in Science Education, 50*, 1739–1761. https://doi.org/10.1007/s11165-018-9751-6.
Potvin, P., & Hasni, A. (2014). Analysis of the decline in interest towards school science and technology from grades 5 through 11. *Journal of Science Education and Technology, 23*, 784–802. https://doi.org/10.1007/s10956-014-9512-x.
Pusey, M., & Pusey, G. (2015) Using Minecraft in the science classroom. *International Journal of Innovation in Science and Mathematics Education, 23*(3), 22–34.
Ritchie, S. M., & Tobin, K. G. (2018). *Eventful learning: Learner emotions*. Brill.
Ritchie, S. M., Hudson, P., Bellocchi, A., Henderson, S., King, D., & Tobin, K. (2016). Evolution of self-reporting methods for identifying discrete emotions in science classrooms. *Cultural Studies of Science Education, 11*(3), 577–599. https://doi.org/10.1007/s11422-014-9607-y.
Ritchie, S. M., Tomas, L., & Tones, M. (2011). Writing stories to enhance scientific literacy. *International Journal of Science Education, 33*(5), 685–707. https://doi.org/10.1080/0950069100372 8039.
Sinatra, G. M., Broughton, S. H., & Lombardi, D. (2014). Emotions in science education. In R. Pekrun & L. Linnenbrink-Garcia (Eds.), *International handbook of emotions in education* (pp. 415–457). Routledge.
Stake, R. E. (2005). Qualitative case studies. In N. K. Denzin & Y. S Lincoln (Eds.), *The SAGE handbook of qualitative inquiry* (pp. 86–109). Sage.
Stets, J. E. (2010). Future direction in the sociology of emotions. *Emotion Review, 2*(3), 265–268.

Thomson, S., De Bortoli, L., Underwood, C., & Schmid, M. (2019). *PISA 2018: Reporting Australia's results. Volume I student performance.* Australian Council for Educational Research.
Tobin, K., Ritchie, S. M., Oakley, J., Mergard, V., & Hudson, P. (2013). Relationships between emotional climate and the fluency of classroom interactions. *Learning Environments Research, 16*, 71–89.
Tomas, L., & Ritchie, S. M. (2012). Positive emotional responses to hybridised writing about a socio-scientific issue. *Research in Science Education, 42*(1), 25–49.
Turner, J. H. (2007). *Human emotions: A sociological theory.* Routledge.
Turner, J. H. (2009). The sociology of emotions: Basic theoretic arguments. *Emotions Review, 1*, 340–354.
Tytler, R. (2007). *Re-imagining science education: Engaging students in science for Australia's future (Australian Council for Educational Research).* ACER Press.
Zembylas, M. (2008). Adult learners' emotions in online learning. *Distance Education, 29*(1), 71–87. https://doi.org/10.1080/01587010802004852.

Senka Henderson (B.Sc. (Hons), M.Sc., Grad Dip Ed, Ph.D., FHEA) is a lecturer in science education at the Queensland University of Technology (QUT), Brisbane, Australia, and teaches in undergraduate and postgraduate courses. Senka's first career included working in pharmacology, biochemistry, and drug discovery, and she has worked in research laboratories investigating new discoveries in carbohydrate chemistry. In her second career as a science educator, she has worked on three ARC (Australian Research Council) science education research projects exploring the emotions of pre-service science teachers in university settings and students in high school science classrooms. She has been involved in other projects related to context-based science education and mindfulness. She is the reviewer and assistant editor of the international journal *Research in Science Education.*

Donna King (B.Sc. (Hons), Grad Dip Ed, M.Sc., Ph.D.) is the National Head of School of Education at Australian Catholic University. Donna's research in science education spans three interconnecting fields: the emotional engagement of students in science in the middle years, using engineering contexts for teaching science and context-based science education. An outcome of this work is the development and implementation of innovative context-based units where teachers have adopted new pedagogical approaches for teaching science. She was part of a team of researchers who completed a project with the Department of Education and Training to establish a STEM Hub for schools where teachers worked with industry partners to connect STEM in the classroom with real-world STEM. She was the lead co-chief editor of the international journal *Research in Science Education.*

Dramatic and Undramatic Emotional Energy: Creating Aesthetic and Emotive Learning Experiences in Science Classrooms

James P. Davis

Abstract The aim of this chapter is to illustrate how everyday mundane actions in science classrooms may be understood as everyday dramatic performances that are most often taken-for-granted and imbued with emotive experiences. To describe these experiences a novel sociological concept of emotional energy (EE) is adopted. EE is an individual and collective experience of togetherness, social cohesion or solidarity arising through embodied performances of situated social practices. It is a way for making visible, cognitive performances and emotive experiences by observing self-coordinating actions, bodily and conversational alignment, mutual bodily entrainment around physical objects, gravitation toward shared ideas, and the fluency of these interactions. In this chapter I make reference to a series of previous studies to describe emotional energy and the notion of everyday dramatic performances. I then describe strategies for educating preservice science teachers about emotional energy as an experience that may be generated and observed in science classrooms through planning and explicit teacher actions. Such awareness by science teachers is important for understanding the influence of emotional engagement in, and dramatic performances of, science learning.

Keywords Dramatic · Emotion · Experience · Performance · Science

Introduction

Drama education is traditionally regarded as learning through the performance of storytelling and fiction (Ødegaard, 2003). More recently the notion *of creative drama* has emerged as a way to describe the application of drama to science education involving emotionally engaging, experiential learning, and integrated curriculum (Ong, Chou, Yang, & Lin, 2020). In this way, drama learning can help with challenging traditional fact-orientated views of science education, focusing toward more contemporary science education objectives that promote science literacy, inquiry

J. P. Davis (✉)
Faculty of Creative Industries, Education and Social Justice, School of Teacher Education and Leadership, Queensland University of Technology, Kelvin Grove Campus, Brisbane, Australia
e-mail: jp.davis@qut.edu.au

P. J White et al. (eds.), *Science and Drama: Contemporary and Creative Approaches to Teaching and Learning*, https://doi.org/10.1007/978-3-030-84401-1_12

skills, and authenticity in learning. The common element between drama and science is the nature of human understanding that relies on metaphorical thinking (cf. Hume, 2007) and features as a foundation for both drama (Henry, 2000) and science (Turnbull, 2000). In drama we perform metaphors and accept them as fiction. In science, we establish socially agreed criteria to sort through our performances of metaphors to tentatively accept some of them as truthful representations of reality. For science, there is a clear fiction-truth conundrum because science metaphors are also producing fiction in the form of representations that are always tentatively grounded in the scientist's imagination (cf. Turnbull, 2000). This is a part of the nature of science, which is commonly misunderstood and misinterpreted in our contemporary times. Clearly, the common ground between drama and science rests with the performance of metaphorical representations by people as actors or scientists, and students of these fields. Grounded on this perspective, in the present chapter I present learning about and through science as a form of drama in everyday situations, where knowing through the metaphorical, embodied performances of science is underpinned by inseparable emotive-cognitive learning experiences (Davis, 2017).

I focus on science learning as a series of everyday performances by children as they engage with *doing science* and *becoming science students*. As children enter the scene of a science classroom their practices and language change, shaping their sense of self, as science students who are learning through doing science. In this sense, they become actors, creating a science identity through the bodily performance of scientific practices and scientific knowledge. These are dramatic performances as children play new roles in doing science via learning events that are experienced through bodily actions and emotion across individual and collective levels of an unfolding reality. I commence by defining emotive experiences in terms of emotional energy, which is a social form of emotional experience distinctively different from discrete emotions such as joy, or frustration. I then describe some examples of creating drama for emotional engagement around science teaching practices, before exploring some techniques for teaching and planning for emotional engagement in science classrooms. These sections draw on recent research conducted into emotions in science education, as well as my own teaching experiences in applying emotions research in preservice science teacher education.

What is Emotional Energy?

Emotional Energy (EE) is conceptualised by Collins (2004) from Durkheim's (1912/1915) notion of collective effervescence. EE is both an individual and a collective experience of togetherness, social cohesion or solidarity arising through embodied performances of situated social practices (Davis & Bellocchi, 2019, 2020). It is empirically and theoretically described in science education research as a way of understanding emotive-cognitive student engagement, evident through self-coordinating actions, bodily and conversational alignment, and mutual entrainment or gravitation toward ideas or physical objects. The relationship between *emotional energy* and *drama* in science education may be viewed through theoretical foundations in Goffman's (1971) sociological work where he describes everyday social interaction as a form of dramaturgy. Dramaturgy is a sociological methodology for interpreting situations of everyday life as being like a drama performance in a front-stage setting, with our private lives concealed in a back-stage setting. In a dramaturgical sense, we are always in-role where some aspects of ourselves are back-staged, and others are front-staged. All social life is a fiction, until we have verified it through interaction, as a social fact (cf. Garfinkel, 1967; Goffman, 1971). A science classroom becomes a scene where some aspects of a child's self are created while other aspects are concealed as back-stage settings. In a successful science lesson, the child performs a new or emergent role, unfolding from moment-to-moment across the lesson. Emotional energy may be thought of as an experience of these dramaturgical interactions and is conceptually grounded partly in Goffman's work (Collins, 2004).

As an experience of front-stage performances emotional energy is often described in ways that are visible and reportable by people in the experience. In science education research these dramatic forms of emotional energy are often characterised with positive, highly visible, and loud verbal expressions. Following Goffman's description of dramaturgy, these characterisations of science classroom experiences are often interpreted as examples of *dramatic* emotional energy (Collins, 2004). Observations of dramatic emotional energy are often supported by self-reports from people as they are typically aware of these highly noticeable front-stage performances and experiences. To illustrate the appearance of dramatic emotional energy, Fig. 1 shows a group of 10-year-old students during a science inquiry activity. This was part of a study I conducted into student engagement during primary school STEM projects in Shanghai, China (Davis et al., 2020). From left to right in Fig. 1, the students are identified with the pseudonyms Wang, Sun, Liu, Qiao and Guo, as they work together to test the fire resistance of a piece of cloth.

As shown in Fig. 1 there is a high degree of mutual bodily alignment in both panels where the students are focused on Sun igniting a piece of cloth. In Panel

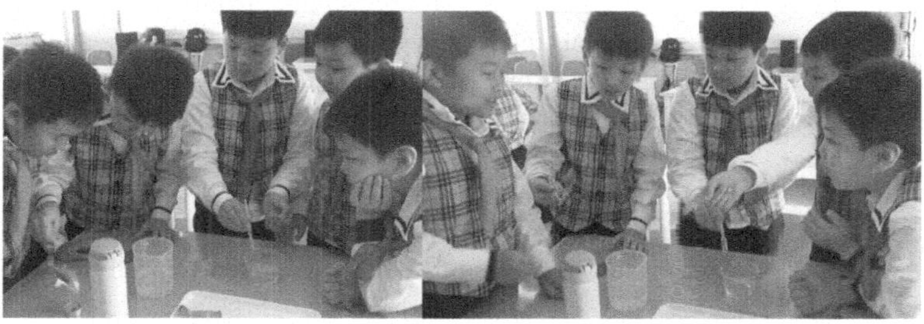

Fig. 1 Panel A, Sun (2nd from left) ignites a cloth. The group is closely huddled with eyes aligned to the flame. Panel B, Sun stands upright, others move back from the flame, maintaining facial alignment toward Sun's right hand. Qiao (2nd from right) points toward the flame [Lab safety as per local protocols]

Fig. 2 Sun's (2nd from left) performance of *how to use tweezers* and Liu's (middle) performance of *how to use the pipette* in a science inquiry activity. Close observation by the group: mutual bodily and facial alignment toward the science instruments [Lab safety as per local protocols]

A all are leaning in closely with face and eye alignment toward Sun's right hand. A moment later, as the flame spreads rapidly up the cloth, Sun pulls his head up, and in the video he makes a "wa" sound consistent with surprise. In Panel B the group have pulled their faces and upper bodies away from the flame, Qiao extends his right arm to point toward the flame, and there is loud chatter about what they had observed. The fluency and self-coordinating character of these actions indicates a high degree of shared focus on the flame, and on each other's actions and reactions. This was a moment of loud excitement, but also intense inter-subjectivity that was described as an experience of high intensity, dramatic emotional energy (Davis et al., 2020). The description of these performed actions illustrates the production of science through embodied, lived experiences that are typically taken-for-granted by people in the experience. If these students were to be asked what was happening here, they would most likely describe their actions in terms of formal scientific methods (cf. Garfinkel, 1967). By describing their performances, we are seeking to understand their performance of science as a social phenomenon that is producing their knowledge of science and self. In this way, these dramatic, front-stage performances may be understood as ways of learning.

In contrast to the experiences portrayed in Fig. 1, intense experiences of emotional energy in science learning situations are quite often unnoticeable to people in the experience (Davis & Bellocchi, 2020). In recent research I have described such experiences in some detail within science classroom situations, as a taken-for-granted form of undramatic emotional energy. In science situations, undramatic emotional energy is observable through self-coordinating actions that are often subtle, involving moments of quietness. Examples include: moments when people become silent at the same time; people making quiet utterances; situations where careful observations are being made; one person making notes in response to another persons' utterances, without any overt arrangement to interact in this way; or, prolonged moments of mutual self-restraint where observers hold themselves back from touching scientific objects (Davis, 2017; Davis & Bellocchi, 2018). A visual example in Fig. 2, shows Wang, Sun, Liu, Qiao and Guo in a quiet moment where Sun and Liu are testing a piece of cloth for water resistance with very little conversation and only slight bodily actions.

The taken-for-granted features of undramatic emotional energy are often evident in everyday situations, but are typically not seen by people in the experience. The mundane nature of everyday actions is evident in the way we humans look *through* our performances as sense-making activities, without looking *at* our performances as our topics of interest (Garfinkel, 1967). If I were to ask these students to recall incidents during the lesson, they are more likely to recall the incident in Fig. 1, rather than Fig. 2, because they would regard the latter events as being mundane, or nothing out of the ordinary. For this reason, emotion in science and science education has traditionally gone unseen, enabling ideas about emotion-free objectivity and rationality to prevail and dominate thinking about these fields of endeavour (Davis & Bellocchi, 2018). This traditional view about emotion-free experiences of doing science may be described metaphorically, as scientists, teachers, or students, being like goldfish, who are looking through the water they swim in, without being able to see the water (Liberman, 2007). In this way, undramatic emotional energy is like water to goldfish. It is an important part of the quieter moments of doing science, which are commonly regarded as undramatic and unemotional because they go unnoticed by people in the experience. This notion of an undramatic performance and experience is an area requiring further consideration in science education studies of emotion. As I will illustrate throughout this chapter, the undramatic may become dramatic, by mobilising teachers' self-awareness of otherwise taken-for-granted experiences.

Creating Drama for Emotional Engagement

Teaching science with active, hands-on learning experiences provides opportunities for learning events that are emotionally engaging. In the emotions in science education literature, examples of these pedagogical strategies influencing emotional experiences include science demonstrations, and various approaches to science inquiry, such as guided and open inquiry. These types of practices may be interpreted from a sociological perspective as rituals performed by teachers and students (Ritchie & Tobin, 2018). In performance and process drama, rituals are viewed as repeated stylised enactments bound by traditional rules (Te Kete Ipurangi, 2020). In science education, natural rituals such as science demonstrations may also be recognised by the rules or patterns of their performance. Rituals in drama are viewed as a means for deepening participants' self-awareness and understanding their actions. In contrast, science teachers tend to look through these rituals to understand them as effective teaching practices, without appreciating how such practices become effective. Studies of classroom rituals as everyday dramatic performances of science have focused on how successful rituals produce high levels of emotional energy focused around science ideas. It is the intensity of emotional energy experienced by science teachers and their students, which attracts the teacher toward the performance of such rituals in future classes (cf. Bellocchi et al., 2018). In this way the emotive experiences of science teaching practices makes repeatability and the formation of natural, everyday, and mundane ritualised practices a possibility.

Science Demonstrations

Science demonstrations are a common pedagogical activity for introducing new topics and stimulating student engagement across all year levels. They are typically structured by introducing an idea about some practical aspect of science, having students predict an outcome, observe the science event, and then explain what is observed–commonly called the Predict-Observe-Explain (POE) strategy. In our teaching at the Queensland University of Technology (QUT), staff and students commonly perform science demonstrations, and for this reason we drew upon the collective experiences of our science education team to analyse these practices from the perspective of emotional engagement (cf. Bellocchi et al., 2018).

I like to think of science demonstrations as a teaching ritual that we perform repeatedly over time. In a study of teacher and student experiences with science demonstrations our team found a diverse range of emotional experiences including discrete emotions such as surprise, and moments of dramatic and undramatic emotional energy. Because emotional energy is about the experience of solidarity with others, people tend to be attracted to practices or rituals that are dramatic and memorable. In the study of science demonstrations by Bellocchi et al. (2018), it was shown how preservice teacher experiences of science demonstrations were connected with repeated performance of these rituals in high school science situations, often months and years after the original performance. In these repeated situations, the performance of science demonstrations was also associated with experiences of solidarity as emotional energy. Bellocchi et al. (2018) described these multi-generational performances of science demonstrations as first order and second order rituals, enabling the dispersion of knowledge through these ritualised performances handed down from teacher to teacher over time. From the perspective as teachers of dramatic and emotive moments in science learning, the possibility of repeating teaching performances to regenerate particular experiences as learning events, is important and should become part of our everyday lesson planning practices. To achieve these idealised practices, science teachers need to become aware of the everyday, mundane actions that are performed as part of the process of science learning.

Science Inquiry

A second area of science teaching where moments of emotional energy are associated with performances are activities involving science inquiry. A study by Ritchie et al. (2013) followed graduate teachers into their first science inquiry teaching experiences involving Extended Experimental Investigations (EEI's) that formerly featured in the

Queensland senior science curriculum. That study revealed the interplay between teacher expectations, teacher experiences in performing EEI's with students, and the influence of teacher emotions on student experiences. The context for that study is situated in prolonged EEI projects over several weeks. These projects involve considerable risk and uncertainty around the outcomes of science inquiry as part of the learning in a senior science assessment context. The art of teaching science in these contexts is dependent on the teacher's interactive skills by observing the particular science performed by students' and asking unscripted and probing questions (cf. Kawalkar & Vijapurkar, 2013). This way of teaching through open science inquiry means both teachers and students became improvisers in their daily performances of science inquiry. In dramatic performance, improvisation involves the invention of drama without preparation or scripting (Te Kete Ipurangi, 2020). From the perspective of everyday performances during open science inquiry activities, science teachers and students are very much improvising, or making up their performances on the run. Such improvisation is itself emotionally demanding and requires science teachers to develop a high level of self-awareness of their own feelings, and their in-the-moment performances and experiences. Teachers' self-awareness is important if they are to shape the unfolding improvisation of interactions with students in science inquiry learning situations.

A further dimension of emotional experience during science inquiry that I led, documents emotional energy during the performance of objectivity by a group of 15 year old Australian students (Davis & Bellocchi, 2018). That performance started with everyday conversation and actions in relation to tea leaves and water in a beaker that were the focus of the inquiry. At the start of the inquiry, students treated these things as everyday objects by touching the water and talking about making a cup of tea. Objectivity became evident as their performances changed by distancing themselves from the water and tea leaves, observing what was happening as the water was heated, and replacing their everyday language with scientific ideas about the actions of the tea leaves. These changes in the way the objects of inquiry were treated by students took place over the duration of a 70 min lesson.

The front-stage where this performance took place was around a workbench in a school science lab. As performers of the science inquiry, students observed the objects of their inquiry, but also each other's performed actions and utterances. At each moment, as students make suggestions about the science, they seek approval for their performances, as a means for changing the fiction of their imagined ideas into accepted facts as scientific statements. They watch one another's actions and utterances, seeking approval for their performances, and reprimanding each other, like an audience of hecklers. This process of dramatic performance is the same process of doing science, where doing science involves the socialisation of fictitious ideas into acceptable facts to represent the reality of their objects of inquiry. This self-coordinating, mutually imposed, unfolding situation also produces an experience of emotional energy that was described as *respect* for the scientific objects. The formation of respect was described in terms of students' subjugating their idiosyncratic senses of self, and collectively engaging in practices that raised the status of the objects as being scientific. This experience and the performance were described as inseparable, intertwined elements of emotion and the performance of thought through actions, toward the achievement of objectivity. That achievement not only transformed a cup of tea into *scientifically* knowable objects, it changed the students through their roles as actors, audience, and mutual hecklers, into *being objective* on the frontstage of a science inquiry.

Teaching and Planning Drama for Emotive Experiences

Having illustrated briefly a couple of examples where drama and emotional experiences of performances may be evident, I will now describe how I teach science education for *dramatic and emotive awareness*. In this section I will outline some key strategies I use in my teaching about emotions in preservice teacher education at QUT. I have applied these strategies across different courses including an elective entitled *Teaching and Learning with Emotions and Relationships in Mind*, a science inquiry pedagogy course and an integrated STEM investigation project. These strategies include raising self-awareness, reflective writing, and role-play to understand emotions in classroom situations.

Self-Awareness of Emotive Experiences

An important element for understanding drama in the performance of science classrooms is to establish self-awareness, which starts with awareness of our own emotive experiences. I tend to start these workshops by having people sit quietly for about 30s, as I ask them to feel the sensation of their bodies. As they sit, I describe for them the nature of this feeling as a raw, unlabelled sensation, and the feeling of being alive. It is what psychologists may call core affect (Russell & Barrett, 1999). These are raw, subjective, biological experiences that can only ever be known by the experiencing person during the experience itself. I then ask people to describe, or label their particular experience of feeling, and this typically elicits quite diverse descriptions in a room of 30–40 people. The diversity of descriptions then becomes a talking point because it illustrates the moment where we transform these raw feelings into emotion, which I describe as feelings objectified with labels. These labels enable us to share an interpretation of our feelings with others. Importantly, the diversity of interpretations where we are all sitting, seemingly doing the same thing, in the same context, highlights the diversity of interpretation and socialisation of biological feelings into something else that we call emotions.

This process of actively becoming self-aware of feelings and slowing down the formation of what we know as emotions is an important, but rarely developed skill for science teachers. It is a practice that can be performed quietly, anywhere, even in the hectic moments of classroom interactions where teachers may sometimes experience a sense of being overwhelmed. Pausing, breathing, sensing, and taking control of the labels we use to make meaning of our sensations can change the drama of a science classroom by changing the direction of a teacher's improvised performance. This in turn changes the performance and experiences of their students.

An additional method that I use in science pedagogy courses where I want to enhance everyday self-reflection skills is the use of heuristic tools that preservice teachers complete each day. Specifically I teach preservice teachers to use a mindfulness (Alexakos et al., 2016), and a speaking and listening heuristic (Powietrzynska & Gangji, 2016) to develop their skills in observing everyday performances of people during their interactions. The level of observation and reflection in these heuristics focuses on fine-grained interactional features that people most typically take-for-granted, such as eye contact, the tone of utterances, and the way we respond to others with our own utterances. These heuristics are very useful for engaging teachers' understandings about the influence their own performances may have in particular interactions with students or peers. Finally, I typically encourage the use of these heuristics when I want preservice teachers to engage in reflective writing, because using the heuristic becomes a means for raising in-the-moment self-awareness of bodily actions, performances and experiences. It enables everyday mundane actions to become visible to teachers, re-positioning mundanity (i.e. an unremarkable experience) to being something extraordinary and worthy of being understood as part of their teaching interactions.

Reflective Writing in Science Inquiry

Supported by heuristics, a further way to raise science teacher awareness of classroom practices as everyday forms of dramatic performance is to engage them in reflective writing. This is achieved by engaging science teachers in performing science inquiry activities in weekly three-hour workshops, reflecting on their performances and emotive experiences, and documenting these reflections in writing. This is a skill that must be explicitly taught, particularly for science teachers who may find writing in first person about subjective experiences quite challenging (cf. Davis, 2019; Korthagen & Vasalos, 2005). I have taught this with a combination of readings about student and teacher's emotional experiences during science inquiry, alongside group discussions during their engagement with and performance of science inquiry activities. Preservice teachers are also supported with readings on how to write about self-reflective events.

I teach reflective writing in science and STEM pedagogy courses where learning involves the performance of a science inquiry project over a number of weeks and reflection on learning experiences. Preservice teachers are encouraged to keep a diary of lab experiences and learning events (cf. Ritchie & Tobin, 2018) as a focus for their reflections. I introduce the need for emotional self-awareness as a way to identify learning events as emotive moments, and I also encouraged them to step back from their projects in the lab, to observe how others are learning and working across the lab environment. This includes taking the time to observe me as their teacher, in the way I engage with different groups to draw out and make explicit their learning experiences. In this way, taken-for-granted actions and performances are made explicit through purposeful observation of, and reflection on, everyday mundane activities.

Role-Playing and Emotive Experiences in Science

Recognising the doing of science in everyday classroom situations as a form of dramatic performance is a means for enhancing preservice teachers' awareness of how abstract notions, such as student engagement, may appear in practice. In my teaching I use the concept of emotional energy (Davis & Bellocchi, 2019) as a social form of student engagement because the performances and experiences of students are visible and hearable through classroom interactions. This observable characteristic of emotional energy makes it a useful tool for teachers during moments of teaching. In my science pedagogy workshops I typically introduce the concept of emotional energy using a social constructivist approach by having students firstly engage in a role-play that we then analyse as a class. In drama learning, role-play requires imagination to identify with some other person to experience and explore a different perspective (Te Kete Ipurangi, 2020).

I have developed the role-play technique for teaching emotional energy with three different roles: Topic Director, Responder 1, and Responder 2. These are played in pairs across two different performances, both led by the Topic Director. I set up the role-play with some interesting object that I bring in from the science lab, which is placed on a table at the front of the class as the focus for role-play interaction. I then ask for four volunteers to perform the roles, and I instruct the remainder of the class to take notes by recording the details about what is said, how things are said, bodily gestures, and facial actions. The role instructions for each actor are shown in Table 1.

The Topic Director starts the conversation and this role is the same in both role-plays. Role-player instructions are in separate envelopes and only the respective role-players see their instructions. With the first role-play involving Responder 1, I am seeking to achieve a highly fluent back and forth interaction, evidencing mutual physical entrainment around the science object as well as a high level of conversational entrainment. Through the interaction a strong sense of shared ideas about the object should be evident, and if fluency can be achieved we could also determine that collectivity in feelings about the topic is also achieved. These elements of bodily movements, ideas, and shared feelings evident through fluency, are the elements comprising intense emotional energy, which I seek to create.

In the second role-play involving Responder 2, I am seeking to achieve disjuncture and a break-down in the interaction. This typically becomes evident from the actions of the Topic Director who fails to achieve a meaningful conversation, and typically breaks from the interaction with laughter, a reddened face from mild embarrassment, or an utterance of disbelief at the actions of Responder 2. This second role-play

Table 1 Role-player instructions

Topic director	Responder 1	Responder 2
Your role is to engage in a conversation. **Focus the conversation on the object** you are given How to do this: • Make statements about the object and keep the conversation focused on the object to the best of your ability • Statements can be in the form of questions, spoken utterances and/or bodily gestures • Each time you make a statement, comment and/or bodily gesture, leave a short pause to let the other person respond • If you find the conversation is progressing well, try to speed the conversation up, by reducing the pause	Your role is to engage in a conversation. **Respond to the other person and focus the conversation on the object** you are given How to do this: • In your responses, make statements about the object and keep the conversation focused on the object to the best of your ability • Statements can be in the form of questions, spoken utterances and/or bodily gestures • Each time you make a statement, comment and/or bodily gesture, leave a short pause to let the other person respond • If you find the conversation is progressing well, try to speed the conversation up by reducing the pause	Your role is to engage in a conversation. **Respond to the other person and *re-direct the conversation*away from the object** you are given How to do this: • In your responses make statements that **counter-act the other person and/or re-direct talk away from the object** • Statements can be in the form of questions, spoken utterances and/or bodily gestures • Try to keep your responses quick and sharp • Try to be matter of fact in your responses. That is, try not to show too much emotion. You want the other person to react toward your bizarre behaviour

highlights divergence in bodily movements as one actor focuses on the science object, while the second actor focuses on anything else except the object. Divergent ideas and a lack of fluency in the interaction tend to be highly visible and hearable.

These two extremes of observed social interaction and different levels of emotional energy intensity then provide content for a class-wide analysis of the interactional performances. The micro-social observations and the discussion around how different actors felt at different moments become important for understanding the conceptual elements of emotional energy and what produces successful performances of social interactions. We are also able to discuss how these concepts apply in science classrooms in the way teachers interact with students. Importantly, we also discuss ways in which teachers should plan and shape interactions to avoid low intensity experiences of emotional energy, such as moments of disjuncture and communication breakdown.

Summary and Future Opportunities

This chapter offers a brief and practical introduction to emotional energy as an experience of embodied performances of different rituals or practices in science classrooms. The ideas in this chapter draw from science education emotions research that highlights a need to change the way we educate science teachers. The proposed change is to raise science teachers' awareness of embodied performances and emotive experiences as dramatic processes in learning science through doing science. While this science education research originates from micro-sociological and phenomenological orientations, through this work I have come to recognise the connectivity between science learning and drama learning. To fully establish this connectivity at a disciplinary level, both science educators and drama educators need to challenge some of the everyday disciplinary phenomenon that they take-for-granted. For both disciplines, creativity, imagination and fiction are important because both fields are grounded in metaphor as a human way for understanding reality. Science educators have long abandoned the notion that science belongs in a laboratory. Science pervades society and our everyday lives. Similarly, I would argue that drama and dramatic performance are not just located on a stage. Dramatic performance is part of our mundane everyday ways of interacting. These everyday mundane sites of interaction are where we perform ideas from our imagination, while seeking to understand ourselves, each other, and our perceptions of reality. This chapter is one step toward making the mundanity of everyday performances a visible phenomenon for science teachers and to raise self-awareness of how drama learning and science learning are interconnected.

References

Alexakos, K., Pride, L. D., Amat, A., Tsetsakos, P., Lee, K. J., Paylor-Smith, C., Zapata, C., Wright, S., & Smith, T. (2016). Mindfulness and discussing: 'Thorny' issues in the classroom. *Cultural Studies of Science Education, 11*, 741–769. https://doi.org/10.1007/s11422-015-9718-0

Bellocchi, A., Davis, J. P., & King, D. T. (2018). Science demonstrations as mediators of interaction ritual chains: Temporal transitions from mundane to intense emotional energy. In S. M. Ritchie & K. G. Tobin (Eds.), *Eventful learning: Learner emotions* (pp. 57–85). Brill.

Collins, R. (2004). *Interaction ritual chains*. Princeton University Press.

Davis, J. P. (2017). Emotions, social beings & ethnomethods: Understanding analogical reasoning in everyday science classrooms. In A. Bellocchi, C. Quigley, & K. Otrel-Cass (Eds.), *Exploring emotions, aesthetics and wellbeing in science education research* (pp. 121–140). Springer.

Davis, J. P. (2019). Preservice teacher learning experiences of entrepreneurial thinking in a STEM investigation. *Entrepreneurship Education, 2*, 1–17. https://doi.org/10.1007/s41959-019-00009-0

Davis, J. P., & Bellocchi, A. (2018). Objectivity, subjectivity, and emotion in school science inquiry. *Journal of Research in Science Teaching, 55*, 1419–1447. https://doi.org/10.1002/tea.21461

Davis, J. P., & Bellocchi, A. (2019). Undramatic emotions in learning: A sociological model. In R. Patulny, A. Bellocchi, S. Khorana, R. Olson, J. McKenzie, & M. Peterie (Eds.), *Emotions in late modernity* (pp. 114–128). Routledge.

Davis, J. P., & Bellocchi, A. (2020). Intensity of emotional energy in situated cultural practices of science education. *Cultural Studies of Science Education*, 15, 359-388. https://doi.org/10.1007/s11422-019-09931-0.

Davis, J. P., Du, J., Jia, H., Qiao, L., Yan, Q., & Chiang, F.-K. (2020). Uniformity, diversity, harmony and emotional energy in a Chinese STEM classroom. *International Journal of STEM Education, 7*, Article 44. https://doi.org/10.1186/s40594-020-00232-5.

Durkheim, E. (1912/1915). *The elementary forms of the religious life* (J. W. Swain, Trans.). Allen & Unwin.

Garfinkel, H. (1967). *Studies in ethnomethodology*. Polity Press.

Goffman, E. (1971). *The presentation of self in everyday life*. Penguin.

Henry, M. (2000). Drama's ways of learning. *Research in Drama Education: THe Journal of Applied Theatre and Performance, 5*, 45–62. https://doi.org/10.1080/135697800114195

Hume, D. (1777/2007). *An enquiry concerning human understanding*. (P. Millican, Ed.). Oxford University Press.

Kawalkar, A., & Vijapurkar, J. (2013). Scaffolding science talk: The role of teachers' questions in the inquiry classroom. *International Journal of Science Education, 35*, 2004–2027. https://doi.org/10.1080/09500693.2011.604684

Korthagen, F., & Vasalos, A. (2005). Levels in reflection: Core reflection as a means to enhance professional growth. *Teachers and Teaching, 11*, 47–71. https://doi.org/10.1080/1354060042000337093

Liberman, K. (2007). *Dialectical practice in Tibetan philosophical culture: An ethnomethodological inquiry into formal reasoning*. Rowman and Littlefield Publishers.

Ødegaard, M. (2003). Dramatic science: A critical view of drama in science education. *Studies in Science Education, 39*, 75–102.

Ong, K- J., Chou, Y- C., Yang, D- Y., & Lin, C- C. . (2020). Creative drama in science education: The effects on situational interest, career interest, and science-related attitudes of science majors and non-science majors. *EURASIA Journal of Mathematics, Science and Technology Education, 16*, em1831. https://doi.org/10.29333/ejmste/115296

Powietrzynska, M., & Gangji, A.-K. H. (2016). "I understand why people need to ease their emotions": Exploring mindfulness and emotions in a conceptual physics classroom of an elementary teacher education program. *Cultural Studies of Science Education, 11*, 693–712. https://doi.org/10.1007/s11422-016-9772-2.

Ritchie, S. M., & Tobin, K. G. (Eds.) (2018). *Eventful learning: Learner emotions*. Brill. https://doi.org/10.1163/9789004377912.

Ritchie, S. M., Tobin, K., Sandhu, M., Sandhu, S., Henderson, S., & Roth, W.-M. (2013). Emotional arousal of beginning physics teachers during extended experimental investigations. *Journal of Research in Science Teaching, 50*, 37–161. https://doi.org/10.1002/tea.21060

Russell, J. A., & Barrett, L. F. (1999). Core affect, prototypical emotional episodes, and other things called *emotion*: Dissecting the elephant. *Journal of Personality and Social Psychology, 76*(5), 805.

Te Kete Ipurangi. (2020). *Arts online: Drama glossary*. The New Zealand Government. Retrieved from https://artsonline.tki.org.nz/Teaching-and-Learning/Pedagogy/Drama/Glossary.

Turnbull, D. (2000). *Masons, tricksters and cartographers: Comparative studies in the sociology of scientific and indigenous knowledge*. Harwood Academic.

James P. Davis (B.Com, B.Nurs. (Hons), GradDipEd, Ph.D., SFHEA) is a Senior Lecturer at the Queensland University of Technology (QUT). In initial teacher education he has taught across science curriculum and pedagogy: foundations of science; integrated STEM; emotions in teaching and learning; and entrepreneurial thinking. His research interests focus on enacted pedagogy and the emotive-cognitive interplay in student engagement and learning from socio-cultural perspectives. He serves as an Associate Editor for *Research in Science Education*, and is a Chief Investigator for Engagement and Learning in the QUT Centre for Inclusive Education. He is also a member of QUT Education's, STEM Education Research Group.

Does Being Positioned in an Expert Scientist Role Enhance 11–13 Year-Old Students' Perceptions of Themselves as Scientists?

Carolyn Julie Swanson

Abstract Student disengagement from school science and the rejection of science-based careers are well known issues. Active inquiry-based approaches are known to enhance learning and engagement in science. This chapter explores whether working in an expert scientist identity through the dramatic inquiry approach Mantle of the Expert (MOTE) supported the students' self-identification of themselves as scientists. This chapter is based on a mixed method study exploring whether MOTE supported the development of twenty-seven 11–13-year-old students' conceptual understandings about buoyancy and stability. Data was generated from assessments, classroom observations, teacher and student interviews and dramatic role conventions. Simple statistical and thematic analysis was undertaken with the findings interpreted through Figured Worlds. Students worked in role as expert scientists re-investigating the sinking of the Ferry Wahine in Wellington Harbour in 1968. An expert scientist identity was explicitly planned and scaffolded. Students mirrored the work of scientists to examine data, conduct science experiments, defend their experimental conclusions and write a report. Student knowledge of the scientific concepts taught improved significantly and scientific terminology usage increased. However, it must be acknowledged that more students did not want to become a scientist but gained a greater awareness of science in society and science-based careers.

Keywords Drama · Identity · Figured worlds · Mantle of the Expert · Science careers

Introduction

Student disengagement from school science and the rejection of science-based careers is a widely acknowledged phenomenon with many contributing factors such as negative perceptions of scientists, gender, identity, the influence of significant others, interest in science, pedagogical approaches used, science achievement and

---

C. J. Swanson (✉)
School of Education, Auckland University of Technology, Auckland, New Zealand
e-mail: carolyn.swanson@aut.ac.nz

P. J White et al. (eds.), *Science and Drama: Contemporary and Creative Approaches to Teaching and Learning*. https://doi.org/10.1007/978-3-030-84401-1_13

school structure (Kennedy et al., 2015; Regan & DeWitt, 2015; Tytler & Osborne, 2012).

The problem is that while many students are positive towards science in general and recognise the value of science to society, being a scientist is "unthinkable" (DeWitt et al., 2013, p. 1055). This may be due to not having enough information about science careers (DeWitt & Archer, 2015; Hendriksen et al., 2015) or the negative discourse surrounding scientists (Bennett & Hogart, 2009; Narayan et al., 2013). Another prominent reason is that students without productive science identities are less likely to pursue science careers or have positive attitudes towards science (DeWitt & Archer, 2015). Therefore, having a strong science identity is a crucial factor in students making "science related choices" (Vincent-Ruz & Schunn, 2018, pp. 8–11) and this may be even more important for girls (Archer & DeWitt, 2015; Vincent-Ruz & Shunn, 2018). It is crucial to create experiences where science and scientists are normalised, as it may have a positive effect on students' perceptions of science, positively impact their science identities and expand their horizons to include science (DeWitt & Archer, 2015).

The critical age range for impacting student decision making about science trajectories is between 10 and 14 years (Cleaves, 2005; Tan et al., 2013). However, when students "are involved in practical, investigative science activities" (Kerr & Murphy, 2012, p. 628) disengagement is less apparent. According to Tytler and Osborne (2012) utilising an active inquiry approach where students learn collaboratively will enhance learning and engagement. A caveat must be given here, it is not enough to only have practical activities, the science being taught must be "connected so [the students] can make sense of their world" (Moeed, & Kaiser, 2018, p. 45). One way that students can connect science to the world is by enlarging the world of the classroom through dramatic inquiry approaches that connect science to society.

The use of drama to support science learning is a growing field (see for example, Abed, 2016; Kolovou & Kim, 2020; McGregor et al., 2019). It has been identified as supporting students' conceptual understandings (Abed, 2016; Braund et al., 2015; Kolovou & Kim, 2020), the Nature of Science (McGregor et al., 2019; Ødegaard, 2003) and science in society (Ødegaard, 2003). The evidence relating to whether drama enhances student attitudes toward science positively is mixed with some authors finding drama enhanced student attitudes (Çokadar & Yılmaz, 2010; Kolovou & Kim, 2020) and others finding positive attitudes slightly decreased (Hendrix et al., 2012). However, integrating the Arts (in this case drama) is not without challenges and training in the approach is needed for optimal teaching and learning (Braund et al., 2015; Turkka et al., 2017).

The dramatic inquiry approach used in this study is Mantle of the Expert (MOTE) (Heathcote & Bolton, 1995). There have been limited research studies exploring the use of Mantle of the Expert in science education (see for example, Aitken & Townsend, 2013; Maxwell et al., 2019; McGregor et al., 2019; Swanson, 2016).

In MOTE students learn curricular subjects in a purposeful manner by working in role as part of an expert team fulfilling a commission for a fictional client within a carefully framed fictional perspective/world (Aitken, 2013; Taylor, 2016). The fictional framing is important as it provides protection (Heathcote & Bolton, 1995) but also "limits and opens possibilities for learning" (Heston, 1994, p. 167). Students enter the drama either through inquiry, drama for learning or expert positioning (Aitken, 2013). While there are at least ten components in MOTE (Aitken, 2013), I suggest that "the company, client and commission are interlinked with, proceed from, and contribute to the positioning of the students as experts" (Swanson, 2016, p. 268). Students agree to take on the role of "someone who is expert at running something" and is committed to meeting the responsibilities associated with the role (Heathcote & Bolton, 1995, pp. 23–28). The expert label is earned through the students embodying their expert role with "increasing conviction, complexity and truth" (O'Neill, 1995, p. viii), as they carry out curricular tasks while fulfilling their commission from the client. It is this taking on of an additional identity–that of an expert scientist–and working in the partial expression of that identity that is the focus of this chapter.

The identity lens of Figured Worlds (Holland et al., 1998) is used as an interpretive lens in this chapter because it has synergies with the imagined and improvised 'as if' worlds of drama (Edmiston, 2003) and has been used in science education to explore, perform and illuminate science identities in the classroom (Tan et al., 2013; Urrieta, 2007a). In this study identity is framed as our personal understandings of who we are, our place in the world, and how we wish to be known and to become (Holland et al., 1998; Schachter & Rich, 2011; Urrieta, 2007b). According to Holland et al. (1998), Figured Worlds are simplified "as if" worlds (p. 49), which are "socially and culturally constructed realm[s] of interpretation in which particular characters and actors are recognized, significance is assigned to certain acts, and particular outcomes are valued over others" (p. 52). The Figured World becomes 'embodied' as the participants play within its framing (pp. 52–53). Over time, the participants move from novice to expert as they engage in significant activities and author their position within the worlds and create their personal identities (pp. 57–60). Key terminology outlined here will be used to interpret the findings.

This chapter will explore the learning that occurred as part of working in an expert scientist identity and discuss whether a nine-week unit shifted students' perceptions of a scientist identity in their future. The research is underpinned by the following research question:

How did Year 7/8 students in this study come to perceive science now and in their future?

Method and Methodology

This study was set within an interpretive worldview (Creswell, 2009), with reality mediated socially and meaning created in dialogue with self and others (Anderson, 2012). This worldview resonates with Heathcote's (1984a) view of drama "filling the spaces between people with meaningful experiences" (p. 97). A form of Participatory Action Research [PAR] was chosen as it had been used in both drama (Cahill, 2006) and science research (Buck et al., 2014). PAR was chosen over a comparative study because MOTE can take considerable time to implement and due to the improvisation nature of MOTE (Heathcote, 1984b) direct comparison would be difficult.

This research draws on data from a mixed-method study based in one semi-rural Year 0–8 'full' primary school in New Zealand. Twenty-seven Year 7/8 students (11–13 year-olds) and their teacher participated in the study. The school was chosen due to the Principal, as the primary gatekeeper to the school (Johnson & Christensen, 2012), being supportive of dramatic approaches. The classroom teacher was a second-year teacher who was familiar with MOTE but less comfortable with science. Full consent and assent were sought and obtained from the participants and their parents (Johnson & Christensen, 2012). The participants were only identified by a pseudonym or their first name as per consent choices. O'Toole and Beckett's (2010, p. 34) drama-based validity criteria of "plausibility, credibility, resonance and transferability" along with Creswell's (2014, p. 201) "validity strategies" were used to outline the rigor of the research.

The unit taught as part of this study was based on a classic MOTE model (Heathcote, 2010) following a format designed by Aitken (personal communication, August 26, 2015) for New Zealand schools. The goal of the research was to explore whether MOTE supported the development of the students' conceptual understandings about buoyancy and stability. Learning was structured within a fictional context with the students positioned as expert scientists who were commissioned to reinvestigate the sinking of the Wahine, which occurred in Wellington Harbour in 1968 and provide answers to a fictional governmental select committee about the factors that contributed to the disaster. The unit included both in-role and out-of-role dramatic approaches and experimentation.

I co-taught the science-based unit, consisting of twenty-two 90-min teaching episodes spread over a nine-week term with the classroom teacher. The approach resonated with the collaborative nature of MOTE (Aitken, 2013; Heathcote & Bolton, 1995) and was a useful way to generate textured insights through being both practitioner and researcher (Menter et al., 2011).

Mixed methods were used for data generation with each method analysed separately, then merged and interpreted with each data set interrogated against each other (Creswell, 2015). Data collected included pre- and post-unit assessments exploring both conceptual and attitudinal data, and audio-taped classroom observations. It was also generated through dramatic role conventions such as Teacher in Role [TIR], Writing in Role and Role on the Wall [ROTW] (Dawson & Lee, 2018; Neelands & Goode, 2000). Semi-structured interviews were conducted. The classroom teacher was interviewed three times, before, after and midway through the unit. Eight students who were purposely chosen in consultation with the classroom teacher for variation in terms of perceived science interest, science ability and gender and who were willing to take part, were interviewed in pairs. A focus group with six of the interviewed students interrogated their interview comments.

For the quantitative arm, answers from the pre- and post-unit assessments were graphed and examined for trends and changes from the pre- to post unit assessments. A two tailed t-test was undertaken to assess whether the students' understanding of the science taught had significantly improved and Cohen's d, a "standardized measure of the magnitude of the effect size" (Field, 2013, p.79) was performed. Effect size can be used to measure student achievement, with an effect size "greater than 0.60" being educationally significant (Timberley et al., 2007, p. 35), and an effect size of 1.0 equivalent to "advancing student achievement by one year" (Hattie, 1999, p, 3).

For the qualitative aspect thematic analysis was carried out with initial codes determined from literature (Creswell, 2009). The student interviews were used as a foundation for the codes, which were supported with observational data from the teaching episodes. Five themes were identified–the main theme pertaining to this chapter was science futures.

In the initial analysis both positioning theory (Harré & van Langenhove, 1999), which explores the effect of position, social force and storyline on discourse, and Figured Worlds (Holland et al., 1998) which can be used to explore "worlds of possibilities" (Urrieta, 2007a, p. 114) were used to examine the data looking at small portions of dialogue and the big picture respectively. However, in this chapter, Figured Worlds is used to explore the development of the students' scientist identities drawing on key terminology as outlined in the introduction.

Findings

This section outlines how an expert scientist identity was created through drama. It describes how the students mirrored the work of scientists in their imagined roles and examines how working in drama supported the students' conceptual understanding of concepts taught. Finally, it discusses how science careers were promoted during the unit and examines the students' possible science trajectories.

Creating an Expert Scientist Identity

MOTE involves deliberate identity work. To an extent the Figured World is authored by the teacher who specifically designs it to maximise curricular learning. However, as soon as the students are invited into the Figured World, it becomes a co-constructed world and takes on a life of its own. The students are in full knowledge that they are agreeing to play in an 'expert' role in an imagined world. There is no compulsion and self-authoring is crucial (Heathcote & Bolton, 1995).

Just like in Holland et al.'s (1998) Figured Worlds, the students in MOTE are inducted into the world through an expert, who knows how the world is configured and what actions are recognised as significant. The imagined world was made significant through sign and word. The sign used included the teacher drawing a ship on the board and creating a visual display in the room with the detritus of a shipwreck to seed the notion of a nautical tragedy. The imagined world was further developed through Heathcote's role convention number 15 in which "objects represent a person's (or in this instance the company's) interests and /values /characters" (Heathcote, 1984c, p. 166), which in this instance was a noticeboard. The items which included the company name Scientific Extreme Event Reconstruction Services [SEERS], a birthday invitation, a letter from a client, a notice about recycling and a weather map were specifically chosen to give the students a glimpse into the company's interests and values. The students were able to identify that SEERS is an "environmentally friendly company" (Josh) that "studies disasters" (Student A) and that they "work well together" (Tom). As they talked about the company, they were making the company embodied.

To further embed the students into the Figured World of the scientific community, students were asked to think about what the company they were exploring would require in terms of office space in an actual world. Students used their prior knowledge of companies and science to co-construct a floor plan for SEERS. They came up with ideas like 'experimenters', 'receptionists', and 'IT people'. The floor plan included general areas like reception, security and advertising as well as specialised areas like the lab room.

Creating resumés or CVs was an important strategy in supporting the students to identify and articulate an identity as someone who was an expert/experienced scientist and a productive member of the company. This involved the students identifying scientist roles and an employment history for themselves within the company. Students positioned themselves as expert scientists, choosing careers like meteorologist and geologists, identifying their commencement date and projects they had been involved in such as helping in the 2011 Christchurch earthquake. They were authoring their identity, identifying with the ethos of the company and giving themselves a legitimate place to stand in the Figured World. They had, according to Hamish, "put on the coat of the person in the business and play[ed] in the role of them".

Mirroring the Work of Scientists

The expert identity was scaffolded by the classroom teacher so that the students' inexpertness was not apparent. One way of doing this was through a collective role (Neelands & Goode, 2000). Students worked in the collaborative expert meteorologist role of Albert to explain about meteorology to me (Teacher in Role [TIR] as one who does not know). Their inexpertness was supported through being given facts about meteorology on small pieces of paper so they could advise the TIR who wanted to learn about meteorology. This activity enculturated the students into the Figured World of the scientist. They communicated their science knowledge to a novice in the Figured World.

Students performed scientific tasks. They formulated research questions on the commission they received from the client to investigate the sinking of the Wahine.

A range of their research questions follows. "Why did the Wahine sink? Why did it turn on its side? What material was it made of? How did that water get in the boat?".

They also analysed articles from the newspapers around the time of the Wahine sinking to find reasons for the disaster, mirroring scientists reading peer-reviewed articles. The students were looking for evidence not opinion. They communicated their research like expert scientists in a team meeting. Brandon said, "It could have been the captain's fault… Cause in our article it said that the public opinion wouldn't allow him to do any more shipping" but Mitchell disagreed quoting his article which reported on the official inquiry stating, "neither the Master nor the Chief Officer of the Wahine was guilty … the charges … were not established". Working in this way enhanced their argumentation skills and criticality.

Students worked in their expert scientist role to conduct experiments. The first two experiments were framed as being from the client. In the first experiment the client wanted the SEERS scientists to re-examine some archival evidence from the fictionalised official Wahine Inquiry. However, after 43 years the experiment was water damaged and the data missing. The archival box also contained a picture of the Wahine on its side lying in Wellington Harbour. Students were asked to use their expertise to reproduce the experiment that was conducted at the inquiry with paper boats and marbles to assist them with their investigation. This experiment supported the students' knowledge about buoyancy, and centre of gravity. Many students used it as evidential evidence in their written report. For example, Ofa one of the lower-level students said, "all the objects on the boat might have weighed a lot like the marbles".

A contributing factor to the sinking of the Wahine was ex-tropical Cyclone Giselle. In the pre-unit assessment only 9/27 (33%) of students were able to provide a basic or partial explanation of cyclone formation. To support the students' understanding of cyclonic winds, the Client asked students to construct a model of a cyclone. This model, in combination with information received from interviewing the TIR as Albert, supported the students' understanding of cyclones in their written reports and assessments. For example, Cameron identified that the cyclonic wind impacted the Wahine's passage into the harbour stating, "while turning the wind and water current changed, now the Wahine was side-on to the entire force of the storm". 15/25 (60%) of the students identified in their reports that not receiving up-to-date weather reports was a factor in the Wahine's sinking. By the end of the unit, 66% (18/27) had some idea with 30% (8/27) having a good or comprehensive understanding. However, it must be noted that students' knowledge was based around Cyclone Giselle and did not transfer completely to tropical cyclones in general with Brooke's comment typical, "we actually focused on Giselle and not on the other ones".

The major way students worked in their expert identity was when they worked in research groups to analyse comments from potential employees around buoyancy and stability. This learning was framed around finding a new employee for the company after Roger, a lackadaisical employee, resigned after a series of suboptimal scientific and professional decisions. The potential employees were asked to do a simple science experiment and write down a scientific explanation for the phenomena. The students in role as expert scientists repeated the experiments and decided which potential employee's statement was the most accurate. Experiments included: floating and sinking objects, making a potato chip/crisp packet float, making potatoes float and exploring free surface effect with plastic containers. Student comment during the experiment time showed they were enjoying the experience. For example, Alicia said, "I'm trying to make this [potato float] … I think it will float. It floats! We hollowed it out and it floats". Belinda continued, "It's got air in it". Students also commented that they liked how the experiments were connected to the wider topic. For example, Lucy said, "we did stuff with fruit [to show] how the Wahine sunk" and Brooke agreed saying, "It sort of showed us how the Wahine sunk". The teacher considered that "when they were doing those experiments, they were [acting like scientists], 'cause they were questioning, they were predicting and recording their results".

To consolidate the learning and help the students justify their decisions, the students were asked to produce a role-play of the job interview with the potential employee they decided had the best understanding of science. The students correctly identified the person with the most accurate science understanding in 5/6 groups. The teacher felt that the learning was consolidated through reflecting on the learning noting "we spent a lot of time on that through the interviews and then … through our report".

The students showed that they were working in full expression of their expert scientist identity when they communicated and defended their findings through writing a report on the sinking of the Wahine. They were asked to back up their statements with experimental data making sure they "had proof". They were supported prior to this by meeting in research groups to collaboratively outline the reasons why the Wahine sank, which they found were related to the weather, general sinking (such as a hole in the boat), density, stability, free surface effect and other aspects like insufficient lifeboats.

All 25 who wrote the report gave a scientific reason for the capsize of the vessel. Twenty-four out of 25 talked about the boat losing buoyancy and/or stability. Fifteen out of the 25 (60%) students supported their reasons with experimental data. Brooke, for example, used the paper boat experiment as a justification for her findings.

> The starboard side started to fill up with water making it more dense and causing it to roll on its starboard side. I found this out because we experimented with paper boats. We placed paper boats in the water and put a marble on one side. This showed that when mass is added to one side the boat will tip to one side.

After examining the reports, the judgement of the teacher was that the quality of their writing was of a higher standard than normal. She noted in conversation with me that "the style and quality of Louise's work was different to what she expected" and mentioned Brandon's report was "quite good" and Kaleb was "doing a really good job". Another teacher Kitt, who worked with reluctant writers, noted they "couldn't wait to get started", not a usual occurrence.

Another aspect that showed the students were working in the embodiment of their expert scientist identity was the development of their terminology from everyday usage to terminology used within the Figured World of science. For example, in week one the students used words like tipped and heavy and by week seven were using words like mass and density. Midway through the unit the classroom teacher confirmed they were more fluent saying, "when they were talking about density and buoyancy, they just say it, the words roll off the tongue". The students themselves identified in the post-unit focus group that early on they "didn't know some of the words" (Lucy), and "had to learn the big ones … like buoyancy" (Cameron) and "density" (Tom). When the written reports were examined, scientific terminology was used by 24/25 students. The words used varied from 'air' to 'weight', 'centre of gravity' with a few alluding to 'free surface effect'. It was also apparent by the end of the unit that more students were using scientific terminology rather than everyday words to describe scientific phenomena.

To add weight to the information contained in the audio transcripts and the written reports, the statistical analysis showed that changes in student understanding between the pre- and post-unit assessments were significant. $t(26) = 7.98, p = 0.0001$, (CI 4.0, 6.8) with Cohen's d effect being 1.6 (Swanson, 2017). The Cohen's d result coupled with the t-test means that MOTE had a large educationally significant impact on science learning in this study.

As students worked in the Figured World of MOTE, they not only took on an expert scientist identity but over time performed recognisable acts like 'real' scientists within the Figured World. Working in an expert identity is different from the classical expression of a Figured World where students come into the world as novices. They worked in partial expression of the expert identity to examine data, conduct science experiments, defend their experimental conclusions, and write a report detailing their findings.

Future Career Choices in Science

As already highlighted in the introduction, one issue in science education is the perception that there are insufficient students progressing to science-based careers. The age of the students in this study (10–14), is recognised as pivotal in setting career trajectories. Therefore, I was interested to see whether working in an expert science identity would have any effect on students' perceptions of themselves as possible scientists. In addition, I wanted to ensure they received pertinent information about both scientists and science careers.

The students' scientist identities were further developed through the dramatic role convention Role on the Wall [ROTW]. ROTW is used to construct a visual representation of a character (Neelands & Goode, 2000). In this case it was used to elicit students' prior knowledge of scientists and the jobs they do (see Swanson, 2016). The students drew an outline of a scientist and in the middle wrote down the characteristics a scientist has and on the outside the tasks they carry out. This supplemented the data collected in the pre- and post-assignment about the tasks scientists carry out. When the data from both sources were analysed, students described scientists as using practical skills (for example, experiment, hypothesise), being theoretical, communicators, and by other activities like 'finding cures'. They described scientists in terms of their thinking skills (imaginative and critical), personality (wacky and witty), academic ability (brainy and well-educated), and having characteristics like "courageous, organised and observant". While the comments the students wrote on their ROTW took a balanced approach, their pictorial drawings positioned scientists as predominantly male (5/6) with crazy hair and wearing lab coats and carrying beakers.

Another way the expert scientist identity was made available for the students to explore was through providing extra information about science careers involving science on a poster on the classroom wall. The careers ranged from biomedical careers like nursing to biosecurity officers, horse trainers and hairdressing. Student comment indicated that they were unfamiliar with many of the careers with Andrew commenting, "I've never seen some of these before". However, it was the inclusion of hairdressing on the poster that piqued discussion about the science in hairdressing. The teacher mentioned that dyeing your hair includes "different peroxides and chemicals [which you] have to balance".

What was apparent at the end of the unit was that the students' perceptions of scientists changed from a stereotypical view of science as seen in the pictures of male scientists with wild hair in ROTW to a more nuanced perception. Taylor admitted with an embarrassed giggle, "last year, I just thought of science as a whole lot of chemicals and people with glasses". While Tom painted a fanciful picture of "scientists as nutty professors creating dinosaurs in dungeons" but went on to state that he now realised "rather than walking around with a lab coats and clipboards … [science] is doing stuff like this [investigation]". Or in Lucy's words, "[Scientists] research and find out quite a lot about the topic and then they figure out what [they] are finding out the answer for and then they experiment".

Students were asked in the assessments what they wanted to be when they grew up. In the pre-assessment 42% (10/24) of the students considered they might have a science career. The six girls were split equally between the life and health sciences. Of the boys, two chose a general science career and one each in engineering and health sciences (veterinarian). When they were surveyed post-unit, the girls remained clustered in the life and health sciences while boys' choices became more specific with one boy moving to the physical sciences.

They were also asked to identify five science careers. Their choices were divided into general science, physical science, life sciences, health science, engineering, careers with science and teaching. In the pre-unit test girls mentioned a career from the health sciences category at double the frequency of boys. Other categories were similar. However, in the post-unit the number of boys mentioning generic careers had dropped from 11 to 7 while the number of boys choosing a career in the physical sciences had increased from 6 to 12. Girl's choices in the health sciences dropped from 18 to 14 while the category remained the same for the boys. Engineering as a career choice doubled for both genders from one to two. Examples of careers that include science also doubled. Science teaching remained the same. While the changes were not significant, student choices became more specific over time.

All interviewed students were asked whether a science career was possible. Tom wanted to be vet, Taylor a doctor, Cameron a career in IT, and Ofa a pre-school teacher who teaches science. Josh wanted a career in sport, which certainly includes science. Jess mentioned three career options in her post-unit assessment that included science: a vet, teacher, or dietitian. Brooke "was not sure" while Lucy stated she "might need to do a bit more work on science because it is something new".

While more students did not want to become scientists, they did gain a greater awareness of science in society and science-based careers. Josh mentioned that working in an expert scientist role in the unit "opened up science more ... I don't think it has shown me more career types. It's just shown me how much science has got to do with jobs and just everyday stuff". The classroom teacher agreed saying "they are so much more aware of what careers are out there that involve science".

Key Findings and Limitations

This section outlines the findings in relation to the research question, how did Year 7/8 students in this study come to perceive science now and in their futures? It focusses on how working within an expert scientist identity within MOTE affected students' perceptions of themselves in science.

Operating in a dramatic Figured World required deliberate planning and scaffolding to introduce the students to the dramatic world and invite them to play within it. Working within the constraints of MOTE meant that careful consideration needed to be given to how the students were positioned within the drama, as they are not positioned as novices as in the Figured Worlds of Holland et al. (1998), but as experts. Dramatic activities were carefully chosen to provide opportunities for the students to author their scientist identity. It was crucial that they have time to transition from students doing science to scientists operating within the full manifestation of the role. Providing a notice board to peruse allowed the students to see what acts were considered 'significant', namely caring for the environment and being a company of team players. Co-constructing a plan for the office and writing personal CVs allowed the students to author a place for themselves within the Figured World of the drama connecting them to a specific place and time. As they played within the world, it was becoming embodied.

Becoming an expert scientist was a valued outcome (Holland et al., 1998), therefore students were provided with opportunities to explore the characteristics of expert scientists and the tasks they undertake through ROTW. The value of the exercise was that it allowed the students to author a collective understanding of scientists and enriching their thinking about possible scientist identities.

It was also clear that working in an expert scientist identity did allow students to play in the Figured World of science and perform the activities that are given significance in science. Within the Figured World of drama they had opportunities not normally available within a classroom and could explore the wider implications of being a scientist and look deeply into the effect of inaccurate science. In their roles as expert scientists they were able to experiment, critique data and defend their science findings in a way that emulated the processes and genre specific aspects of science (Vosniadou, 2012), albeit in an imagined but legitimate domain.

Working in an expert scientist identity through the dramatic inquiry approach MOTE positively impacted upon the students' learning as their knowledge of the science concepts taught improved significantly and scientific terminology usage increased. The judgement of the teacher was that the students not only wanted to share their scientific findings, but the quality of their writing was of a higher standard than normal. However, it must be acknowledged that from the study design and findings it is not possible to say whether MOTE was more effective than any other teaching approach in enhancing conceptual understanding.

One possible reason for students not progressing into science-based careers is not being supplied with critical information about science careers (DeWitt & Archer, 2015) when they are formulating possible careers for themselves. Therefore, information on science-based careers was provided in the form of a poster. It was apparent from student comment that they had limited knowledge about the range of science-based careers available. However, when the data from the pre- and post-unit assessments, and student interviews was analysed, there was no discernible difference in students' science trajectories apart from the possible career ideas being more specific in a couple of instances. Working in an expert scientist identity is not a quick solution to solving the perceived problem of insufficient numbers of students undertaking higher science degrees or becoming scientists.

As this was not a longitudinal study there was no way to assess whether or not these students on the cusp of high school carried on with their science education. However, what they did leave the study with was an understanding of the interconnectedness of science which Moeed and Kaiser (2018) suggest is an important component in encouraging students to remain on a science trajectory. They gained a greater understanding of how science was connected to society, and science-based careers.

There were a few limitations in research design that may have impacted the findings. The research took place in one classroom over one term. This meant that the data was often too small for substantive statistical analysis. In addition, the discontinuous nature of this study may have impacted student engagement and scientist identity formation, as the momentum was not able to be sustained. It would be useful to undertake a longitudinal study to see if working in an expert scientist identity had an impact over higher education in science and career choices in the future.

References

Abed, O. H. (2016). Drama-based science teaching and its effect on students' understanding of Science concepts and their attitudes towards science understanding. *International Education Studies, 9*(10), 163–173. https://doi.org/10.5539/ies.v9n10p163.

Aitken, V. (2013). Dorothy Heathcote's mantle of the expert approach to teaching and learning: A brief Introduction. *Connecting curriculum: Linking learning* (pp. 34–56). New Zealand Centre for Educational Research.

Aitken, V., & Townsend, L. (2013). Searching for the truth/s: Exploring enviro-ethics. *Connecting curriculum, linking learning* (pp. 57–82). New Zealand Centre for Educational Research.

Anderson, M. (2012). *Masterclass in drama education: Transforming teaching and learning.* Continuum.

Archer, L., & DeWitt, J. (2015). Science aspirations and gender identity: Lessons from the ASPIRES project. In E. K. Hendriksen, J. Dillon & J. Ryder (Eds.),*Understanding student participation and choice in science and technology education.* (pp. 89–102). https://doi.org/10.1007/978-94-007-7794-4_6.

Bennett, J., & Hogarth, S. (2009). Would you want to talk to a scientist at a party? High school students' attitudes to school science and to science. *International Journal of Science Education, 31*(14). https://doi.org/10.1080/09500690802425581.

Braund, M., Moodley, T., Ekron, C., & Ahmed, Z. (2015). Crossing the border: Science student teachers using role-play in grade 7. *Journal of Research in Mathematics, Science and Technology Education, 19*(2), 107–117. https://doi.org/10.1080/10288457.2015.1016711

Buck, G., Cool, K. L., Quigley, C. F., Prince, P., & Lucas, Y. (2014). Seeking to improve African American girls' attitudes towards science: A participatory action research project. *The Elementary School Journal, 114*(3), 431–453. https://doi.org/10.1086/674419

Cahill, H. (2006). Research acts: Using the drama workshop as a site for conducting participatory action research. *NJ (Drama Australia Journal), 30*(2), 7–26. Retrieved from A+ Education database.

Cleaves, A. (2005). The formation of science choices in secondary school. *International Journal of Science Education, 27*(4), 471–486. https://doi.org/10.1080/0950069042000323746.

Çokadar, H., & Yılmaz, G. (2010). Teaching ecosystems and matter cycles with creative drama activities. *Journal of Science Education and Technology, 19*(1), 80–89. https://doi.org/10.1007/s10956-009-9181-3.

Creswell, J. W. (2009). *Research design: Qualitative, quantitative, and mixed method* approaches (3rd ed.). Sage.

Creswell, J. W. (2014). *Research design: Qualitative, quantitative and mixed methods approaches: International student edition* (4th ed.). Sage.

Creswell, J. W. (2015). *A concise introduction to mixed methods research.* Sage.

Dawson, K. & Lee, B.K. (2018). *Drama-based pedagogy: Activating learning across the curriculum.* Intellect.

DeWitt, J., & Archer, L. (2015). Who aspires to a science career? A comparison of survey responses from primary and secondary school students. *International Journal of Science Education, 37*(13), 2170–2192. https://doi.org/10.1080/09500693.2015.1071899.

DeWitt, J., Osborne, J., Archer, L., Dillon, J., Willis, B., & Wong, B. (2013). Young children's aspirations in science: The unequivocal, the uncertain and the unthinkable. *International Journal of Science Education, 35*(6), 103701063. https://doi.org/10.1080/09500693.2011.608197.

Edmiston, B. (2003). What's my position? Role, frame, and positioning when using process drama. *Research in Drama Education: The Journal of Applied Theatre and Performance, 8*(2), 221–230. https://doi.org/10.1080/13569780308334.

Field, A. (2013). *Discovering statistics using IBM SPSS statistics* (4th ed.). Sage.

Harré, R., & van Langenhove, L. (Eds.). (1999). *Positioning theory: Moral contexts of intentional action.* Blackwell.

Hattie, J. (1999). *Influences on student learning.* [Inaugural Lecture: Professor of Education]. University of Auckland. Retrieved from https://cdn.auckland.ac.nz/assets/education/about/research/documents/influences-on-student-learning.pdf.

Heathcote, D. (1984a). Drama and learning. In L. Johnson & C. O'Neil (Eds.). *Dorothy Heathcote: Collected writings on education and drama.* (pp. 90–102). Northwestern University Press.

Heathcote, D. (1984b). Improvisation. In L. Johnson & C. O'Neil (Eds.). *Dorothy Heathcote: Collected writings on education and drama.* (pp. 44–48). Western University Press.

Heathcote, D. (1984c). Signs and portents. In L. Johnson & C. O'Neil (Eds.). *Dorothy Heathcote: Collected writings on education and drama.* (pp. 160–169). Western University Press.

Heathcote, D. (2010). Internal coherence—a factor for consideration in teaching to learn. A paper to explain the interior planning and outer praxis when a drama element is used in working in 'Mantle of the Expert' mode with students in a middle school in Victoria Canada May 2009. *Drama in Education, 26*(1), 24–66. Retrieved from http://www.mantleoftheexpert.com/wp-content/uploads/2010/03/Jan-101.pdf.

Heathcote, D., & Bolton, G. (1995). *Drama for learning: Dorothy Heathcote's mantle of the expert approach to education.* Heinemann.

Hendriksen, E. K., Dillon, J., & Giuseppe, P. (2015). Improving participation in science and technology higher education: Ways forward. In E. K. Henriksen, J. Dillon & J. Ryder (Eds.),*Understanding science participation and choice in science and technology education* (pp. 367–379). https://doi.org/10.1007/978-94-007-7793-4_2.

Hendrix, R., Eick, C., & Shannon, D. (2012). The integration of creative drama in an inquiry-based elementary program: The effect on student attitudes and conceptual learning. *Journal of Science Teacher Education, 23,* 823–846. https://doi.org/10.1007/s10972-012-9292-1.

Heston, S. (1994). *The construction of an archive and the presentation of philosophical, epistemological and methodological issues relating to Dorothy Heathcote's drama in education approach* (Doctoral dissertation). Lancaster University, UK. Retrieved from http://www.partnership.mmu.ac.uk/drama/HESTON/Phd.pdf.

Holland, D. C., Lachicotte, W., Jr., Skinner, D., & Cain, C. (1998). *Identity and agency in cultural worlds.* Harvard University Press.

Johnson, B., & Christensen, L. (2012). *Educational research: Quantitative, qualitative, and mixed methods* (4th ed.). Sage.

Kennedy, I., Smith, P., & Sexton, S. S. (2015). Ensuring New Zealand's future prosperity: A professional learning development initiative to bridge the gap between theory and practice. *Science Education International, 26*(1), 42–55. https://files.eric.ed.gov/fulltext/EJ1056452.pdf.

Kerr, K., & Murphy, C. (2012). *Children's attitudes to primary science.* In B. J. Fraser, K. Tobin & C. J. McRobbie (Eds.), *Second international handbook of science education: Part 1* (pp. 627–650). Retrieved from http://www.ebrary.com/corp.

Kolovou, M., & Kim, N. (2020). Effects of implementing an integrated drama inquiry model in science class. *The Journal of Educational Research.* https://doi.org/10.1080/00220671.2020.1771673.

Maxwell, D., Pillatt, T., Edwards, L., & Newman, R. (2019). Applying design fiction in primary schools to explore environmental challenges. *The Design Journal, 22*(sup 1), 1481–1497. https://doi.org/10.1080/14606925.2019.1594972.

McGregor, D., Baskerville, D., Anderson, D., & Duggan, A. (2019). Examining the use of drama to develop epistemological understanding about the nature of science: a collective case from experience in New Zealand and England, *International Journal of Science Education, Part B,* 9(2), 171–194. https://doi.org/10.1080/21548455.2019.1585994.

Menter, I. J., Elliot, D., Hulme, M., Lewin, J., & Lowden, K. (2011). *Guide to practitioner research in education.* [Ebrary Reader version]. Sage.

Moeed, A., & Kaiser, S. (2018). Staying in science: Can understanding of the disciplinary connectedness of science help? *Teaching Science, 64*(2), 45–50.

Narayan, R., Park, S., Peker, D., & Suh, J. (2013). Students' images of scientists and doing science: An international comparison study. *Eurasia Journal of Mathematics, Science &Technology Education,* 9(2), 115–129. https://doi.org/10.12973/eurasia.2013.923a.

Neelands, J., & Goode, T. (2000). *Structuring drama work: A handbook of available forms in drama and theatre.* University Press.

Ødegaard, M. (2003). Dramatic science. A critical review of drama in science education. *Studies in Science Education, 39*(1), 75–101. https://doi.org/10.1080/03057260308560196.

O'Neill, C. (1995). Foreword. *Drama for learning: Dorothy Heathcote's mantle of the expert approach to education* (pp. vii–x). Heinemann.

O'Toole, J., & Beckett, D. (2010). *Educational research: Creative thinking & doing.* Oxford University Press.

Regan, E., & DeWitt, J. (2015). Attitudes, interest and factors influencing STEM enrolment behaviour: An overview of relevant literature. In E. Henriksen, J. Dillon, & J. Ryder (Eds.), *Understanding student participation and choice in science and technology education* (pp. 63–88). Springer.

Schachter, E. P., & Rich, Y. (2011). Identity education: A conceptual framework for educational researchers and practitioners. *Educational Psychologist, 46*(4), 222–238. https://doi.org/10.1080/00461520.2011.614509.

Swanson, C. (2016). *Positioned as expert scientists: Learning science through mantle of the expert at Years 7/8*. (Doctoral dissertation). The University of Waikato, Hamilton, New Zealand. Retrieved from http://hdl.handle.net/10289/9974.

Swanson, C. (2017). Fictional others: Expanding the possible through interactions with the fictional.*NJ Drama Australia Journal, 41*. https://doi.org/10.1080/14452294.2017.1329678.

Tan, E., Calabrese Barton, A., Kang, H., & O'Neill, T. (2013). Desiring a career in STEM-related fields: How middle school girls articulate and negotiate identities-in-practice in science. *Journal of Research in Science Teaching, 50*(10), 1143–1179. https://doi.org/10.1002/tea.21123.

Taylor, T. (2016). *A beginner's guide to mantle of the expert: A transformative approach to education*. Singular.

Timperley, H., Wilson, A., Barrar, H., & Fung, I. (2007). *Teacher professional learning and development: Best evidence synthesis iteration [BES]*. Retrieved from http://www.educationcounts.govt.nz/data/assets/pdf_file/0017/16901/TPLandDBESentireWeb.pdf.

Turkka, J., Haatainen, O., & Aksela, M. (2017). Integrating art into science education: a survey of science teachers' practices. *International Journal of Science Education, 39(10)*, 1403–1419. https://doi.org/10.1080/09500693.2017.1333656.

Tytler, R., & Osborne, J. (2012). Student attitudes and aspirations towards science. In B. Fraser, K. Tobin & C. McRobbie (Eds.), *Springer international handbook of education: Second international handbook of science* (pp. 597–625). Retrieved from http //www.ebrary.com/corp.

Urrieta, L. (2007a). Figured worlds and education: An introduction to the special Issue. *Urban Review, 39*(2), 107–116. https://doi.org/10.1007/s11256-007-0051-0.

Urrieta, L. (2007b). Identity production in figured worlds: How some Mexican Americans become Chicana/o activist educators. *Urban Review, 39*(2), 117–144. doi:https://doi.org/10.1007/s11256-007-0050-1.

Vincent-Ruz, P., & Shunn, C. (2018). The nature of science identity and its role as a driver of student choices. *International Journal of STEM Education, 5*(48), 1–12. https://doi.org/10.1186/s40594-018-0140-5.

Vosniadou, S. (2012). Reframing the classical approach to conceptual change: Preconceptions, misconceptions and synthetic models. In B. Fraser, K. Tobin, & C. McRobbie (Eds.), *Second international handbook of science education* (pp. 119–130). Retrieved from www.ebrary.com/corp.

Carrie Swanson (DMLS, GDipT, PGDEd, (Dis.), Ph.D.) is a lecturer in Initial Teacher Education at Auckland University of Technology, New Zealand. Her current speciality areas of teaching are primary science education and human development, but she has taught in drama education. Her research interests include science education, dramatic inquiry approaches including Mantle of the Expert, integrated teaching approaches, action research and identity lenses.

Stories from History: More Authentic Ways of Thinking Through Acting and Talking About Science

Debra McGregor

Abstract This chapter will discuss and demonstrate how it is possible to introduce young people to scientists' life stories and draw on particular events or incidents that can inspire them to think more deeply about science. Noteworthy moments from historical scientific stories are dramatised in various ways to engage learners to consider these scientific happenings from different perspectives. The learning activities, adopted and adapted from established theatrical strategies, purposely oriented learners to think about the lives and work of scientists from varied viewpoints. Immersed and positioned differently in a range of historical contexts to work in-role enabled learners to consider science from alternate perspectives. This provided not only an historical dimension to learning about science, but many of the narratives the learners were introduced to offered insights about socio-cultural influences determining what and how scientists investigated in the past. Learners working in-role, in participatory ways, considering issues that faced scientists in the past, can inform and shape age-appropriate inquiry tasks. Drawn from a series of action research projects carried out in schools across the UK, ways that different theatrical strategies have been developed and trialled in classrooms to engage young people with stories form history are carefully described so that others might apply these approaches.

Keywords Historical stories · Inquiry · Drama pedagogy · Participatory learning

Introduction

Texts used to teach scientific or technological concepts are often presented in an authoritarian style to learners that is impersonal and abstract. As such, science is portrayed as "a set of objective truths" and "absolute realities" (Avraamidou & Osborne, 2009, pp. 1684). Introducing ideas in a more concrete and personalised way, however, is possible through a variety of texts, including more fictional approaches that provide stories or narratives that capture students' imaginations (Davies &

D. McGregor (✉)
Oxford Brookes University, Oxford, UK
e-mail: dmcgregor@brookes.ac.uk

P. J White et al. (eds.), *Science and Drama: Contemporary and Creative Approaches to Teaching and Learning*, https://doi.org/10.1007/978-3-030-84401-1_14

McGregor, 2017). Fictional stories in science can even be used to introduce factual content through the introduction of both conceptual ideas and narratives from "historical events" (McGregor & Precious, 2015, pp. 113–114). McCullagh et al., (2010, p. 23), have described how this can make science "more relevant" to children's lives. That is, fictional texts can offer a medium by which science can be communicated in a more meaningful, relevant and accessible way (Avraamidou & Osborne, 2009). Utilising storytelling that places scientists' lives and work into a 'real' context for children through theatrical approaches (McGregor & Precious, 2015, pp. 103–125) can therefore be seen as a personalised way of thinking about science. Stories can help develop an understanding (Mutonyi, 2016), respect and even appreciation of the discipline's historical roots, through narrations taking account of culture, time, place and situation (Meadows, 1987). The hybridisation of narratives of a fictional nature, brought to life through forms of drama or theatrical techniques (Neelands & Goode, 2015) whilst weaving in references to factual information can, therefore, be used as a culturally epistemic tool that naturalistically enables learners (even young children) to begin to understand more complex scientific ideas (Mutonyi, 2016; Anastasiou et al., 2015). This chapter therefore describes how stories or narratives that emerge from well-known scientists' lives can be transformed into immersive activities applying drama conventions to engage young learners epistemically in thinking about science from a more personalised and empathetic perspective.

Using Historical Stories

Traditionally historical stories have provided context when teaching older students or even science undergraduates. Narratives, for example, about how Galileo Galilee famously dropped two cannonballs of different weights from the top of the Leaning Tower of Pisa to illustrate principles in physics about rates of falling bodies (Van Cleave, 2004, p. 72), contribute to ways of thinking and teaching about the derivation of ideas in science. Engaging learners in thinking about the life and work of scientists, who might seem remote and disconnected from their daily lives, can enhance their appreciation of them as real people and even enable them to begin to understand why and how they come up with the discoveries they did.

Relating stories in scientists' lives to the science being learned can extend opportunities for students to more directly relate to part of their family, social context, their personal experiences, or even their struggles as children growing up without parents or being born poor. Considering stories relating to scientists and/or inventors they have heard of, for example, Isaac Newton or Alexander Graham Bell, can be drawn on to introduce how they might have felt or behaved when they were younger. Newton, for example, grew up without a father, he was lonely as a child with no siblings, brought up by his grandmother (Meadows, 1987, p. 70) on the family farm. Inspired by his surroundings he used tools (like his penknife) and made what he could from natural materials he found on the farm (twigs, leaves, hessian, twine etc.) to make things (like small model boats when he younger and larger 4-wheeled carts when he was a youth). Alexander Graham Bell had a deaf mother and his father taught children with hearing difficulties. Growing up aware of the way that communication could be extended beyond just talking face-to-face, he was intrigued when he played the piano. His mother, by the way that strings, pedals and other devices could generate vibrations (and sounds), could detect his playing through an ear trumpet.

Connecting Drama and Historical Stories

Drama can offer quite unique opportunities for students to learn in a participatory (O'Toole, 1992) and enjoyable (McGregor & Precious, 2015) way. The wide variety of ways that drama conventions (Farmer, 2011; Neelands & Goode, 2015) can be adapted and adopted to nurture curiosity and confidence in acting and talking about science can engage students in memorable learning experiences. It can bring to life a "story from the past" (McGregor & Precious, 2015, pp. 171–227) and conjure students' empathy to better understand historical situations and the context of important scientific discoveries. It can be used to develop an excitement and curiosity about science, but most of all it can introduce to students what it means to *be* a scientist (McGregor, 2012; McGregor et al., 2014). Used in sequence, drama conventions can enable students to be successively immersed in thinking about a scientist, their life and work, in a progressive and deepening way. One of the more unique ways developed by McGregor & Precious (2015) to introduce someone from the past, is to use a monologue, that is a mini-speech, given by a teacher or a student in-role, as if spoken by the scientist themselves. These short speeches can orally present listeners with interesting and thought-provoking information from the life of the scientist that is intriguing to think about.

It is also possible to engage students in enacting a script of an historical play. A teacher or actor can dress up and speak like an historical scientist from the past (Stagg, 2019). The teacher, actor or student that has spoken in-role as the scientist can then be 'hot seated' (Farmer, 2011, p. 28). This technique places the speaker 'in-role' to be questioned about being that character. The person in the hot seat usually sits on a chair and is somewhat in the spotlight when questioned by the others in class. The person in-role might be an expert scientist or a character from history with a point of view that others are asking questions of, or, about. A useful convention to adopt subsequently, is the tableau (Farmer, 2011, p. 67; Neelands & Goode, 2015), whereby an idea or concept is represented as a 3D still illustration. Engaging with stories about scientists, students can adopt still or statue-like poses to represent the different attributes, scientific practices (Osborne, 2014), or 'actions' (McGregor & Precious, 2015, p. 59) that are undertaken to solve a problem or make a discovery. A very frequently used drama convention is that of miming movement (Farmer, 2011, p. 79; Neelands & Goode, 2015), whereby participants are asked to mime or demonstrate with bodily movements, actions or gestures how they think something works.

This physicalisation of ideas to enact a 'scientific process' or even 'events' (McGregor & Precious, 2015, p. 47) in a scientist's life could include 'freezing' actions to illustrate an important moment (Farmer, 2011, p. 63), a particular occasion or significant happening. Besides these adaptations of drama conventions, students can be positioned (Swanson, 2016) to work in-role as a particular scientist (McGregor & Precious, 2015, pp. 121–124) and are given the props or equipment reminiscent of that scientist's life and era, to work with to respond to a query, open question or inquiry. The ways that the story of a technological scientist's life and work has informed a series of related drama activities and inquiry-like activities is that of William Harbutt (the creator of plasticine) described below.

Bringing a Story to Life Through Drama: An Example of a Technological Scientist William Harbutt

In a West-Midlands school in the UK, a nine-year-old child, dressed in a suit and tie, read a script as if he was William Harbutt (the creator of plasticine). His classmates listened as he read the monologue entitled 'The most perfect material' (McGregor & Precious, 2015, p. 219). He read the script as if he were the scientist reflecting on his life and work. He indicated how as a boy he was creative in making things and then later as an art teacher using clay, he became frustrated that it dried too quickly before his students could manipulate it into its final form. The class listened to the story in absorbed concentration as he explained how the clay set hard too rapidly between classes and students could not refine the shapes and produce the detailed figurines he wanted them to create. He decided to create a clay that was more flexible for longer for his students to work on their sculptures. He asked a friend, an old soldier who knew about mixing different substances to make explosives to help him! He wasn't interested in blowing things up, but how combinations and proportions of different substances (some of which are used for explosives) like calcium salts, oils and petroleum jelly can be stirred together to make a flexible clay. He explained how he recorded what *mix* of salt, oil, jelly and clay was added together each time he tried a new combination. Each new *recipe* was stirred together in the gigantic vat in his cellar, which he then squashed or compressed to squeeze out the liquid and produce a more malleable modelling clay. His story included talking about how his own three children, who loved to play with the putty-like material, suggested he could make it more interesting and appealing by adding colour. The story finishes with an explanation of Olive, the eldest daughter, who when much older, helped him run the family business selling plasticine in England and even overseas in New York.

Listening to a speaker in the role as the scientist reflecting back on his life provided the context for subsequent dramatised activities (Neelands & Goode, 2015) related to Harbutt's life and work. The teacher then invited the class to question the boy, that is 'hot seat' (Farmer, 2011, p. 28) him and ask questions to find out more about him. The students became very involved in this approach and asked many questions.

What emerged was that they thought he was *artistic, had good ideas, cared about his students, was persistent, observant, tried different things systematically, wrote things down* and *was thoughtful.* The hot seating discussion enabled the students to appreciate that a scientist or inventor can be more than the often held view of a white coated, spectacled and balding man (Chambers, 1983). Many other 'stories' like this (McGregor & Precious, 2015, pp. 103–126, 171–227)) can be developed to support students in realising how scientists' childhoods or aspects of their family lives may resonate with their own.

After being introduced to William Harbutt's story through a monologue the class in the Midlands school produced a series of tableaus depicting the ways he was scientific. An example of one of these, produced by a group of eight- and nine-year-olds is shown in Fig. 1. To elicit from the students what their view of their pose conveys, tapping them on the shoulder invites them to 'narrate' (Neelands & Goode, 2015, p. 129) what they are embodying. One student declares he is communicating *anxiety*, indicating how it is not easy to produce the 'perfect' plasticine; another is gesturing as if *mixing* substances; another is *moulding the clay*; a girl is thinking about her list of *ingredients for the clay*; the kneeling boy represents the old soldier *helping Harbutt mix* his clay together; whilst the final child is *overseeing* the process, that is watching over and monitoring the drying and production of the plasticine.

Fig. 1 Example of a student generated tableau, a still 3D depiction of the scientific traits they understood Harbutt demonstrated through their reflection on the monologue

Extending the Dramatisation of an Historical Story to Present a Related Inquiry Opportunity to the Students

Having some appreciation of the scientific approach Harbutt, as a technological scientist, applied in his life (after listening to the monologue, enacting episodes of his life in a tableau), the students are then invited to decide how to create their own modelling clay. It was made clear to the students that they were then in-role as Harbutt, applying his traits as a scientist. They (the students) had to decide within their groups what their ingredients were to be, the proportion in which they would mix them as well as clarify the other practical steps they would take to create their doughy clay. They needed to measure how much of each ingredient to add; consider how to mix them; think about what to add for colouring; record the outcome of the mixture; and test the mixture to assess its use as a modelling clay, its malleability but retention of a firm shape, consequently creating their own unique plasticine. As such they created their own original methods for making their clay.

The following illustrations (Fig. 2a, b) shows some of the steps the eight and nine-year-olds took to produce their Harbutt clay.

Once made, the students tested the 'clays' in many creative ways. They laid out them out on a long roll of wallpaper (see Fig. 3a), the length of the classroom, labelled their clay (see Fig. 3b), and then tore off a small piece to test each in turn. The tests they devised included rolling the dough like a sausage and see which was the longest before it broke; gently pulling it from both ends to explore its length before it cracked to see how stretchy it was; dropping it from a certain height to see if it remained intact when it hit the floor, and making a standing arch with it and timing how long it held its shape (as illustrated in Fig. 4).

Finally, they gave each clay a score to decide which was most malleable, retained its colour and could effectively be shaped as required, all features that resonated Harbutt's plasticine.

This kind of activity provided the opportunity for the students to discuss their ideas about what were the most useful ingredients to include, so they were beginning

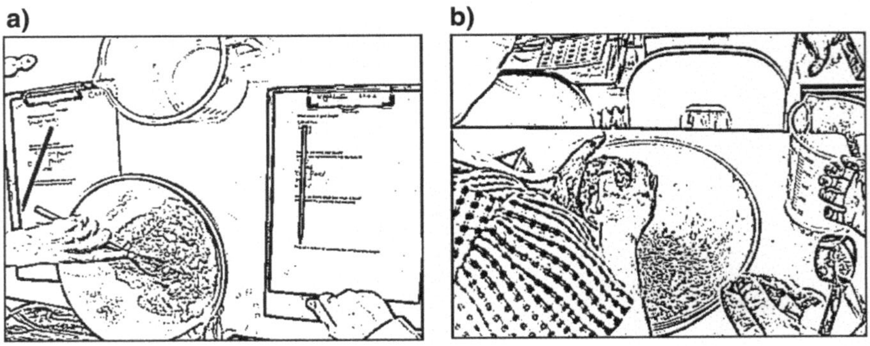

Fig. 2 **a**, **b** Steps in the process of making the Harbutt clay

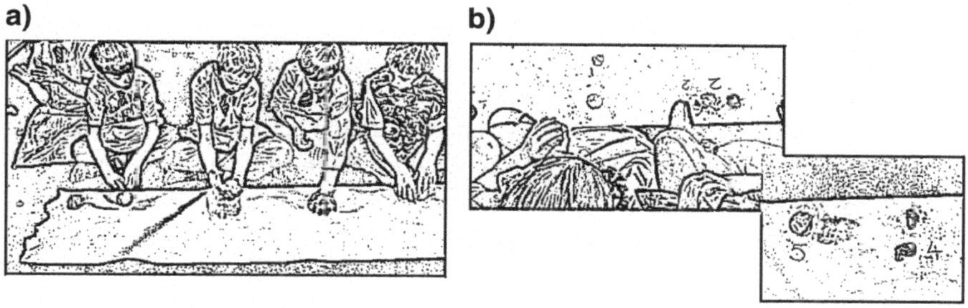

Fig. 3 **a** Lining up the different clays; **b** Labelling the clays ready for testing

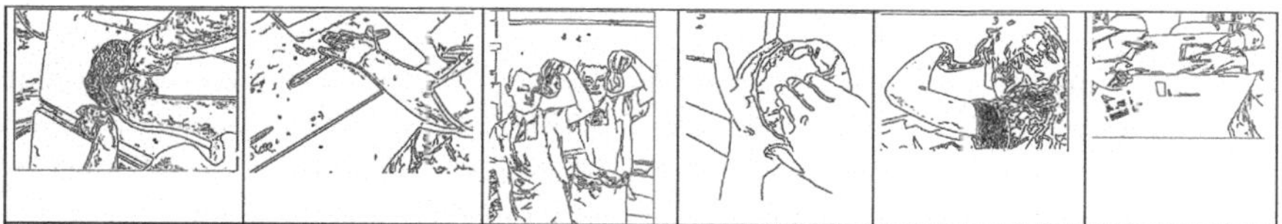

How squashy is it?	How well does it roll into sausage?	How long before one breaks and drops to floor?	How long does it stay in shape?	How long before the bridge breaks?	How far can it be stretched before it tears?

Fig. 4 Illustrations of some of the ways the students tested the clay material

to discuss materials and their properties. They questioned each other about the final recipes that contributed to each different clay, so they began to compare whether there were particular ingredients that worked better to make it more flexible or colourful, for example. The students appeared to easily relate to being scientists and to imagine how they needed to act when working as such. They used a wider range of scientific vocabulary when discussing which ingredients for the different clays were best for retaining their shape, staying smooth, keeping their colour and not drying out too quickly.

This kind of dramatised approach engaging students to work in-role, to not only think about the lives of scientists (Baskerville & Anderson, 2015; McGregor, 2016; McGregor et al., 2019), but to also participate in an inquiry related to their context, aids their appreciation of what a scientist is and what a scientist does. They grasp how scientists had real families, siblings (e.g. Wedgewood who had ten siblings), pet dogs (e.g. Darwin who had beagles), and may have had disabled family members (e.g. Graham Bell's deaf mother), or needed to earn money to buy them food (e.g. Mary Anning beach combing with her pet dog, finding fossils on the Lyme Regis coastline and selling them to visiting Edwardian tourists).

However, another tack to grab their attention is to select a scientist they may not have heard of, but whose contribution they encounter in everyday life, e.g. the creator of Velcro (Mestral) or plasticine (Harbutt), or rubber for car and bicycle tyres (Dunlop). Introducing lesser well-known people through monologues spoken as ‘mystery scientists’ (McGregor & Precious, 2015, p. 103) can deeply engage students in thinking about the social, historical and cultural aspects of scientists’ lives.

Integrating more personal information into the drama activities for the students when they are acting in-role, inspires them, enables them to relate science to their everyday lives, and even helps them appreciate that they *can* work scientifically, better understand science, and even become a scientist or technologist in the future and make a real difference to the world (McGregor et al., 2019).

Engaging in and with Scientific Inquiries

Extending thinking about and introducing simple open tasks related to different aspects of a scientist’s life and work can offer illustrations and/or opportunities for students to practice the development of various inquiry capabilities as indicated in Table 1. Posing an investigative question or query that the students could work on in-role as a scientist provides additional motivation because they recognise the context of working the way they are. They can better appreciate the purpose of their working out an answer or response to questions that make sense to them, as outlined in Table 1. Drawing on the William Harbutt story, for example, to set-up an inquiry, the focused questions were: i. What is the best recipe for a modelling clay?; and ii. What kinds of tests can you carry out to find out which is the best clay? Answering each of these open questions resonates with aspects of Harbutt’s work, in the way he experimented with different ingredients to create the best modelling clay for his students.

To engage in more authentic inquiry situations, extending the opportunity for students to each work out an original solution, requires good quality ‘open’ questions to be posed. Framing the Harbutt activity with given instructions (pre-determined by the teacher) to make a clay and then test it, constrains the extent to which the students can devise their own approaches to creating a material and then testing its properties. Posing a question such as those outlined above *(i. What is the best recipe for a modelling clay? ii. What kinds of tests can you carry out to find out which is the best clay?)* intimates that there are potentially multiple ways of making and testing the clay. This approach certainly provided authentic inquiry learning opportunities for the students.

Adopting the stories of scientists’ lives and work is possible, as indicated in Table 1 to frame inquiries. There are varied ways of generating opportunities for students to work in-role as a particular scientist, especially if the context of their work is introduced through the use of drama conventions described earlier that serve to immerse the learners in a rich and meaningful context generating purpose in the inquiry activity. To complement this, appropriate resources can be introduced to increase the authenticity and challenge in responding to the open inquiry questions (as indicated in Table 1). Practical experimental work experiences, then, can be designed to consolidate the development of a range of inquiry and scientific practices (Osborne, 2014). Trialling these approaches with teachers through professional development approaches can result in many of them adapting their teaching to include a range of freshly developed inquiry opportunities for students (McGregor, 2012).

Table 1 Example of the ways that monologues (McGregor & Precious, 2015, pp.171–227, 156–160) can be used to emphasise a part of the scientists' story, which can then inform a related inquiry activity

Scientist	Aspect of their 'story' that can be drawn upon	Inquiry practices that could be emphasised	Materials for the suggested inquiry	Possible inquiry question
Marianne North (monologue: McGregor & Precious, 2015, p. 209)	Travelling overseas to exotic places, with no instant camera, drawings and paintings had to be made to capture details of the observations of the different species seen	Thinking creatively to explain things Making comparisons looking for simple patterns Using observations and measurements to draw conclusions	Coarse paper (black and white), charcoal, chalks, water colours, pencil, different plants, range of fruits, vegetables & a variety of seeds (the more exotic the better)	Which seeds could be produced by which plants? Why do you think that?
Joseph Shivers (monologue: McGregor & Precious, 2015, p. 207)	After repeated experiments eventually Shivers created a much sought after material, lycra	Testing ideas using observation and measurement Trying things out (deciding about evidence, equipment and materials) Generate a fair test Check observations and measurements by repeating them when appropriate Make comparisons and identify simple patterns Use observations and measurements to draw conclusions	Range of materials: jersey, cotton, wool, paper, card, bubble wrap Bowl, bucket, jug, spoons, tape measure, ruler, water	What have you found out about the materials? How could you find out which material is best for an Olympic sport?

(continued)

Table 1 (continued)

Scientist	Aspect of their 'story' that can be drawn upon	Inquiry practices that could be emphasised	Materials for the suggested inquiry	Possible inquiry question
Ben Franklin (monologue: McGregor & Precious, 2015, p. 211)	The experience of an electrical charge when flying his kite with his father	Thinking creatively to explain things Asking questions Testing ideas using observation and measurement Trying things out (deciding about evidence, equipment and materials) Making comparisons and identifying simple patterns Deciding whether the conclusions made are supported by evidence	Balloons: different sizes, shapes, colours Different fabrics, spoons, tissue, space blanket, foil, feathers, comb, rulers, paperclips, rocks, metals Plant sprayers and water (to create different humidities)	What have you found out about static electricity? How does static electricity effect our everyday lives?
Michael Faraday (monologue: McGregor & Precious, 2015, p. 213)	Abstract conceptualising to explain the behaviour of an electrical circuit	Testing ideas using observation and measurement Trying things out (deciding about evidence, equipment and materials) Designing a fair test Checking observations and measurements by repeating them when appropriate Making comparisons and identifying simple patterns Using observations and measurements to draw conclusions	Range of different batteries (3), battery tester, additional wires, foil sheet, cork, rubber, shells, graphite pencils (9B, 6B, 3B, HB), charcoal, wire wool, range of 3 rocks (including haematite), old fork, small squares of metal (nickel, cobalt, copper, iron)	How many different ways can you make a circuit that works? What (variables) might make a difference to the circuit?

Teacher Reflections on the Use of Drama and Historical Stories to Teach About Science

Teachers report that using stories from history is one of the most challenging aspects of applying drama techniques to engage students in thinking (and inquiring) in/about science (McGregor & Precious, 2015). However, after practicing the strategies and becoming more familiar with the way they could be implemented, teachers noticed that some of their students were more able to: (i) design their own bespoke investigations; (ii) develop more informed justifications and reasons to explain their findings; and (iii). discuss their ideas using a wider scientific vocabulary. For example, in developing an inquiry from the story of Joseph Shivers (McGregor & Precious, 2015, p. 207), who created Lycra to make clothing more stretchy for different kinds of (exercising or fashion) purposes. One teacher reported how she adapted this story as a prompt for thinking about properties of materials. She gave her students a range of different materials from which to explore which was the most waterproof. After asking an open question, *Which material is the most waterproof?*, the students offered lots of ideas. So, the teacher provided them with limited equipment, such as 10cm^2 squares of different materials (cotton, plastic sheet, bubble wrap, foil), two beakers, teaspoon, jug, bowl, pegs, timer, and in groups they had to devise a way of finding out which material was most waterproof. When asked, *Why are you doing it that way?*, they would reply with a variety of reasons, indicating, *because of this and that and the other...* What the teacher noted was that at first they found it difficult to really focus on one variable at a time. However, after more practice responding to open questions and devising their own inquiries, the pupils became more confident and it was easy for the teacher to just say, *Right, now investigate what would happen if...* and they could! Other teachers reported how this approach to inquiries even helped students with learning disabilities, i.e. those with Special Educational Needs (SEN). The stories really captured their imagination and resulted in focused work because they (the SEN students) were interested and really motivated to find a solution to the open questions posed (McGregor & Precious, 2015).

Students' Reflections on the Use of Drama and Historical Stories to Learn Science

Often students will say they enjoy engaging in drama activities in science lessons, but they highlight how it is fun for different reasons. They may suggest: *It is better than sitting down*; *You don't have to write*; *It is more interesting*; *You can act it out*; and because it is active they appreciate it is participatory learning and not just sitting down writing. Interestingly, they recognise the inclusivity and the social aspect of drama, providing an opportunity for everyone. Quite a significant proportion describe how it explains more through observing others acting out interpretations. It provides a visual conveyance or alternate presentation of ideas. Some students even feel it helps them empathise with others' situations, people or objects more, because they say: *It is like you are the person and then you learn more.*

This is the kind of connection Neelands (2002) describes as understanding ourselves in relation to what goes on around us. For students it is an opportunity for them to share what they know with their peers and make sense of it through talking and deciding how to 'perform', 'move', 'make sounds', 'act like a…' etc. One student described how after enacting the process of being a living creature in the sea and eventually becoming fossilised on the seashore where Mary Anning might have found her, said drama was good because, *We know what it feels like.* Incorporating student's prior-knowledge and embedding activities into their everyday experiences can help them start to see connections between science and their close surroundings, which it is argued acts as a motivating factor (Koballa & Glynn, 2007). Typical comments from students, after being involved in a Mattie Knight related inquiry (McGregor, 2016) for example, include: *I like science … but what we did last […] made me enjoy science even more and in, normal science lessons, we mixed and solved stuff, but in this you were using your imagination as well and creativity which I really like.* However, one particular student, indicated how it is the immersion in the activity that it not only motivates but focuses and extends the attention paid to science, because *you are concentrating more […] you have put yourself in the person's shoes.*

The Impact of Using Stories in Drama to Improve Inquiry Learning in Science and the Nature of Science

Engaging in inquiry practices that are contextually informed by a story from history (McGregor, 2016) can provide learning opportunities to rehearse and enact scientific practices (Osborne, 2014) as well as develop an appreciation of the nature of science (McGregor et al., 2014, 2019). The kinds of scientific practices that it is possible for students to engage in through a dramatised inquiry includes them being able to ask questions of each other as they devise their group approach to an investigation. They might, as in the example provided earlier, ask questions of another student in-role as a scientist. They can contribute to class discussion about what kinds of scientific questions they could all answer. During the Harbutt activity, they had the opportunity to construct explanations from their observations, for example, the students striving to explain why one clay is more malleable than another. In the clay-making activity, they used the evidence from all the groups to begin to discuss and argue which ingredients (oil, flour, water, salt, food colouring etc.) were important in which combinations and proportions to create a flexible clay. As indicated in Table 1, they devised various ways to test the clays so that they could consider the evidence they generated to evaluate which clay material was best for creating models or objects.

Engaging in inquiries, students themselves could recognise the development of different kinds of scientific practices. Asked about the kinds of things they had done after they had participated in dramatic inquiries, they were able to indicate a range

Table 2 A summary of student perspectives reflecting on the kinds of scientific practices they experienced in a dramatic inquiry experience (from McGregor et al., 2019)

Inquiry practice reflected upon	% pupils' responses
Asking questions	100
Thinking of new ideas	100
Testing ideas	96
Observing how things change	68
Comparing things	86
Using evidence to make conclusions	91
Using scientific words	100
Making decisions like a scientist	86
Thinking like a scientist	96
Acting like a scientist	96
Being a scientist	96

of practices that they had taken part in. For example, after involvement in the Mattie Knight related inquiry, students reflected on their activities as summarised in Table 2.

Conclusion

The various ways that stories can be drawn upon, adopted, adapted and brought to life through engaging in different kinds of drama strategies has been illustrated in this chapter. Drama conventions can be successfully employed to help students think about science and enact being a scientist in-role. Immersing learners in thinking about what scientists' lives were like, how they came up with their ideas and drew on resources available to them at that time, through taking up roles of well known (or not so well known) characters from the past can provide informed insights into historical events that shape how the world is today.

Dramatising historical stories engages students in a different way of thinking and learning about science. It can enable learners to enter an imagined world, an 'as if' context (Anderson, 2004), where they can envision themselves to be someone or something different. This new perspective means that they are positioned differently and so find it easier to consider what science is from an alternate view-point and ask different kinds of questions. Working in-role creates different kinds of spaces for learning where there are more opportunities to discuss and negotiate what something means as well as how it might feel. Through the dramatised inquiry activities students can engage in processes by which unknown scientific outcomes can emerge. Through this lived experience, where in-role they make decisions about what to do, they can authentically experience the true nature of science (McGregor et al., 2014, 2019).

Providing students with these kind of inventive opportunities to engage with the development of scientific ideas through dramatising episodes of historical stories, teachers generate opportunities for learners to explore alternate forms of expression that can enrich reading, writing (McGregor & Precious, 2015, p. 166) and scientific literacy. As Neelands (2002, p. 8) indicates this kind of learning activity means that students with "a broader range of ability [have] access to an expressive form". It, therefore, appears that bringing alive stories from history can offer a range of scientific insights for learners, as well as enhance their literacy, motivate them to solve problems and even inspire them to want to become scientists.

Acknowledgements The author would like to acknowledge that aspects of this project were supported by the Primary Science Teachers Trust (https://pstt.org.uk/resources/cpd-units/dramatic-science).

References

Anastasiou, L., Kostaras, N., Kyritsis, E., & Kostaras, A. (2015). The construction of scientific knowledge at an early age: Two crucial factors. *Creative Education, 6*(2), 262–272.

Anderson, C. (2004). Learning in 'as if' worlds: Cognition in drama in education. *Theory into Practice, 43*(4), 281–286.

Avraamidou, L., & Osborne, J. (2009). The role of narrative in communicating science. *International Journal of Science Education, 31*(12), 1683–1707. https://doi.org/10.1080/09500690802380695

Baskerville, D., & Anderson, D. (2015). Investing in the pretend: A drama inquiry process to support learning about the nature of science. *Research Information for Teachers*, 50–58.

Chambers, D. W. (1983). Stereotypic images of the scientist: The Draw-a-Scientist test. *Science Education, 67*(2), 255–265.

Davies, D., & McGregor, D. (2017). *Creative teaching in primary science* (2nd ed.). Routledge.

Farmer, D. (2011). *Learning through drama in the primary years*. David Farmer.

Koballa, T. R., & Glynn, S. M. (2007). Attitudinal and motivational constructs in science learning. In S. K. Abell & N. G. Lederman (Eds.), *Handbook of research on science education*. Lawrence Erlbaum Associates.

McCullagh, J., Walsh, G., & Greenwood, J. (2010). Books and stories in children's science. *Primary Science, 111*, 21–24.

McGregor, D. (2012). Dramatizing science learning: Findings from a pilot study to reinvigorate elementary science pedagogy to five-to-seven year olds. *International Journal of Science Education, 34*(8), 1145–1165.

McGregor, D. (2016). Using drama within a STEM context to develop enquiry skills and appreciate being a scientist! *Journal of Emergent Science, 12*, 16–24.

McGregor, D., Anderson, D., Baskerville, D., & Gain, P. (2014). How does drama support learning about the nature of science: Contrasting narratives from the UK and NZ. In C. P. Constantinou, N. Papadouris, & A. Hadjigeorgiou (Eds.), *E-Book Proceedings of the ESERA 2013 Conference: Science education research for evidence-based teaching and coherence in learning: Part 6: Nature of science: History, philosophy and sociology of science* (pp. 22–33). European Science Education Research Association. Retrieved from http://www.esera.org/media/esera2013/Debra_McGregor_16Feb2014.pdf.

McGregor, D., Baskerville, D., Anderson, D., & Duggan, A. (2019). Examining the use of drama to develop epistemological understanding about the nature of science: A collective case from experience in New Zealand and England. *International Journal of Science Education, Part B, 9*(2), 171–194.

McGregor, D., & Precious, W. (2015). *Dramatic science. Inspired ideas for teaching science using drama ages 5–11*. Routledge.
Meadows, J. (1987). *The history of scientific discovery*. Phaidon.
Mutonyi, H. (2016). Stories, proverbs, and anecdotes as scaffolds for learning science concepts. *Journal of Research in Science Teaching, 53*(6), 943–971.
Neelands, J. (2002). *Making sense of drama: A guide to classroom practice*. Heinemann.
Neelands, J., & Goode, T. (2015). *Structuring drama work: 100 key conventions for theatre and drama*. Cambridge University Press.
Osborne, J. (2014). Teaching scientific practices: Meeting the challenge of change. *Journal of Science Teacher Education, 25*(2), 177–196.
O'Toole, J. (1992). *The process of drama. Negotiating art and meaning*. Routledge.
Stagg, B. (2019). Meeting Linnaeus: Improving comprehension of biological classification and attitudes to plants using drama in primary science education. *Research in Science Technological Education, 37*(1), 15–35.
Swanson, C. (2016). *Positioned as expert scientists: Learning science through mantle of the expert at years 7/8* (Doctoral dissertation). Waikato University, Hamilton, New Zealand. Retrieved from http://hdl.handle.net/10289/9974.
Van Cleave, J. (2004). *Scientists through the ages*. Wiley & Sons.

Debra McGregor (B.Sc, M.Sc, Ph.D., PGCE) Debra's research work centres on teaching and learning processes that span subject disciplines especially in STEM and STEAM contexts. As a former primary and secondary teacher and now researcher, her educational experiences inform the multiple ways that she considers, examines and articulates pedagogical enactments and the relational nature of teaching and learning. She believes the application of drama conventions to enable an 'as if', imagined, or even Figured World, can be a powerful pedagogical approach that extends ways for learners to act in-role as scientists, engage in thinking like a scientist, and even aspire to become a scientist.

Australian Women in Science: A Model for a Research-Based Theatre Project in Secondary School Classrooms

Richard Johnson Sallis and Jane Bird

Abstract This chapter reports on research conducted by the authors drawn from both primary and secondary data sources based on two theatre productions performed by students at the University of Melbourne. Both performances featured characters based on female alumni, and current students from a range of faculties were amongst the cast members. The researchers found that a common element of both productions was the emphasis on women in science and, developmental and performance processes akin to those used in research-based theatre. Research-based theatre is an increasingly popular form of research presentation in the social sciences. As the name suggests, it draws on research data and findings and shares them via performance. The authors argue that research-based theatre may provide a useful model for a factually-based ensemble (or group-devised) performance in senior secondary schools. The model they share is intended for use by teachers and their students in a range of secondary school learning areas, including Science and Humanities. The authors have developed their model based in part on their work with pre-service and in-service teachers who wish to embed drama pedagogies into their planning and teaching of non-drama subjects. When developing the model shared in this chapter, the authors also drew on their collective knowledge and experience in the areas of cross-curricular applications of drama and research-based theatre play-making processes.

Keywords Research-based theatre · Women in science · Drama · Secondary schools

Setting the Scene

There remains a need to attract more young women into Science/STEM in senior secondary schools and undergraduate Science programs (Dasgupta & Stout, 2014;

R. J. Sallis (✉) · J. Bird
Melbourne Graduate School of Education, University of Melbourne, Melbourne, Australia
e-mail: sallis@unimelb.edu.au

J. Bird
e-mail: jmbird@unimelb.edu.au

P. J White et al. (eds.), *Science and Drama: Contemporary and Creative Approaches to Teaching and Learning*, https://doi.org/10.1007/978-3-030-84401-1_15

Kang et al., 2019; Salmon, 2015). Research has also been conducted which indicates that drama pedagogy is an effective method by which to explore issues of gender politics in schools and in the wider community (Hatton, 2020; Hatton & Sallis, 2019). Drama, as a subject in schools and as pedagogical method, is well-known for providing students with an opportunity to explore their own life experiences and those of others, past and present (O'Toole et al., 2009). In the teaching of drama there is an understanding that whilst it is a subject or arts area in its own right, and often taught by trained drama teachers, drama is also a teaching method that can be utilised in other learning areas in schools (Sinclair et al., 2017). In this chapter it is the latter approach that is taken by the authors. We demonstrate how drama pedagogy can be used to create and present a research-based theatre performance about Australian female scientists to promote Science and STEM/STEAM/STEMM to female students and others in educational institutions. Importantly, the model we have devised can be utilised by teachers with their students in non-drama subject areas such as Science, English and the Humanities.

Some of the key terms we use in this chapter are as follows:

Ensemble performance (or group-devised play):

> … developed and presented by a group … who shares the responsibility for the devising of it… [with a] focus on the expressive skills/performance skills of the performers rather than on elaborate props, set and costumes … The performers … have the artistic control over the material they will be presenting. Individuals within the ensemble collaborate by sharing responsibility for the devising and the presentation of the performance (Bird & Sallis, 2019, p. 16).

A group who devises and presents an ensemble performance is known as an 'ensemble group' or an 'ensemble'.

Eclectic theatre: The specific performance style for the ensemble performance we propose in the model we share later in the chapter is 'eclectic', drawing on methods of research-based theatre. Many eclectic theatre performances draw on both ancient theatre traditions and contemporary theatre styles (including non-Western theatre styles). The focus is more on the actor than on elaborate staging. However, multimedia (pre-recorded video sequences, soundtracks, data-shows and so on) is increasingly being used in eclectic theatre performances. Some other elements of eclectic performances include: the use of tableaux (frozen images); the use of a non-linear plot structure (using flashbacks and flashforwards); transformation of object, place and time; one actor playing more than one character; the use of a narrator or commentator; use of symbol; use of song, dance, puppetry, mask and mime; use of physical movements and gestures (Bird & Sallis, 2019, pp. 14–15).

Research-based theatre: In this chapter we use the terms 'research-based performance' and 'performed research' interchangeably as do many of our contemporaries (Belliveau & Lea, 2011, 2016). Research-based theatre draws on the traditions of ethnodrama and ethnotheatre (Saldaña, 2005, 2011), ethnographic performance (Denzin, 2003) and performed research (Belliveau & Lea, 2011). For Belliveau and Lea research-based theatre "present[s] research in a way that is compelling and captivating, connecting with viewers on imaginative and intellectual levels" (Belliveau & Lea, 2016, cover). One who co-devises or writes a research-based play is often referred to as being an 'ethnodramatist' (Carter & Sallis, 2019). Being research-based, the performance utilises both primary and secondary sources, that is, document analysis or desk-top research, as well as interviews and other appropriate qualitative data collection instruments. Some performed research plays draw from quantitative data as well. This performance style aims to heighten an audience's awareness "that what they are looking at and listening to is revelatory and truthful" (Soans, 2012, p. 19).

Immersive theatre: According to internationally renowned immersive theatre director Jason Warren, a performance becomes a piece of 'immersive theatre' when those who devise and perform it come up with ways to engage directly with the audience when it is being performed (Warren, 2017).

Improvising: is "to act spontaneously with little or no prior preparation … the actor(s) think and work 'on their feet' to make up what happens" (Bird & Sallis, 2019, p. 22).

Scripting: is "the process of developing a script … it does not necessarily need to refer to a traditional script (incorporating dialogue etc.). It can refer to a detailed description of scenes presented in the order in which they will be performed" (Bird & Sallis, 2019, p. 23).

The Research Project

In the research project which informed the writing of this chapter the authors sought to investigate how a play about Australian female scientists devised and performed by university students could provide a useful model for a similar project in senior secondary schools. Case study (Yin, 2013) and reflective practitioner (Schön, 1984) methods were employed. Our findings revealed that, whilst not being referred to as such by their creators, two recent group-devised performances at the University of Melbourne (UoM) about female academics appeared to use processes akin to those applied in research-based theatre. As the name suggests, this form of performance draws on research data and findings and shares them via performance. Often the dialogue spoken by the characters is derived from (previously spoken or written) words of the real people on which the play is based. It is well documented that research-based plays can raise awareness, change attitudes and affect change, including in schools around issues of gender bias and subject selection (Ackroyd & O'Toole, 2010; Belliveau & Lea, 2011; Hatton, 2020; Hatton & Sallis, 2019). This knowledge, combined with the research we conducted into the two university performances, informs the model we have devised.

Research-based performances were a part of both our doctoral studies at the UoM. Jane studied the work of female academics at our university and wrote a research-based interactive performance piece which was performed by professional actors at a Women in Leadership conference. Richard worked with students in a senior secondary school Drama class to present a researched-based play he had written which reported on his research at the school into the gendered nature of teaching and learning in drama. In one of our most recent projects we draw on interviews with over fifty drama teachers with thirty or more years of professional experience to write a script (along with university colleague Kelly McConville) for the play *The Greatest Love of All* (Bird et al., 2019b). The play was subsequently performed at a national drama educators' conference in 2018.

Performances of Women in Science

The University of Melbourne has a long and proud history of student performance. Its many clubs, colleges and societies regularly stage theatrical productions. The budget for these performances is usually modest and requires the student theatre groups to do more with less. A noticeable trend in recent years has been an increase in original theatre works written and produced by current students and/or alumni. Whilst the university has two dedicated theatres which are specifically intended for student performances, other university spaces are also utilised, including classrooms, lecture theatres and outdoor spaces. This research project featured two recent student-led productions. Common characteristics of the two performances included:

- They were devised by/with the students;
- The storyline of the plays was informed by research conducted into the real-life characters on which they were based;
- The intended audience was primarily other students at the university;
- The plays, at least in part, celebrated the achievements of female scientists, including those who had studied at the University of Melbourne;
- The plays highlighted the past challenges of studying and practicing science and medicine as a female, and invited the audience to consider what may, or may not, have changed over time.

Play 1: *A Journey Through Time*

The first of the two student-led performances we studied was, *A Journey Through Time.*

> March 8, 2019, and the University of Melbourne (UoM) is celebrating International Women's Day. An audience has gathered in the grounds of the university to watch an immersive theatre piece. The initiative of [the then] Associate Director, Workplace relations, Chancellery, this short immersive theatre performance will be performed numerous times during the day. Entitled, 'A Journey Through Time', its focus is 'a celebration of women at the University of Melbourne' from 1883 to the present. The characters are based on successful female alumni. As the performance begins, a guide leads the audience through a tunnel opening under the South Lawn into a living exhibition where they engage with actors portraying notable female students from the university. As a collective, the audience witness a 'journey through time' featuring key women alumni who achieved greatness, especially in the sciences whilst at the university and/or after graduating. (Sallis, Fieldnotes 03/19)

The real-life characters featured included UoM alumni Dr. Lucy Meredith Bryce who in the early 1900s helped to establish the blood transformation service in Australia, and Kate Isobel Campbell noted for discovering the cause of blindness in premature babies.

Play 2: *Ida–The Musical*

In 2018, the Union House Theatre (UHT) at the University of Melbourne staged *Ida–The Musical,* a reimagined interpretation of the 1994 UoM student production, *The Princess Ida Parlour*, directed by Petra Kalive. According to Kalive's program notes:

> Women were first admitted to study at [the University of Melbourne] in 1880 as a result of a Royal Commission in Victoria's Education System. In 1883 [UoM student] Bella Guerin became the first female graduate from an Australian University. (Kalive, 2018, para. 3)

Our research into the background of the play uncovered that the 'Princess Ida Parlour' was formed in 1888 by a group of UoM female students and operated until 1925. It derived its (ironical) name from a popular Gilbert and Sullivan light opera of the time, *Princess Ida.* The opera which satirises feminism and women's education, tells the story of a princes who founds a women's-only university, only to have her attempts thwarted by men. At the university, the 'Parlour' was a physical space (two rooms on the upper story of the east side of the Law Quad) where women studying at the university could meet to discuss and debate issues and, as necessary, mount campaigns against the male-dominated hegemony of the university. Debating was particularly encouraged in the 'parlour' because it was seen as a highly useful tool with which to counter the obstructions of the male gatekeepers whose decisions directly affected the academic lives of the female students. One of the scenes in the play is entitled 'Penniless Blue Stockings'. As explained in the performance, in the eighteenth century, intellectual males were known to wear blue stockings. The early female students were branded by their male peers as being 'Blue Stockings', the implication being that they were *masquerading* as male students or academics. In turn the female students, like those in the wider feminist movement, came to own the term and used it to refer to early female tertiary students. As the play details, in 1887, twenty-five years after the opening of the UoM Medical School, female students were finally permitted to study there, with the first woman doctor graduating in 1891. Significantly, historical data such as this, which informed the development of the play were selected based on how they aligned with the themes the writer(s) wished to explore. For example, the way in which the female and male university characters (students and lecturers) were depicted in the play was shaped by the choices of those who devised, directed and performed it. In both of the performance pieces it became evident that even in a play based on factual information, a degree of subjectivity and artistic licence will affect what ends up on stage.

The Plays: Content, Themes and Style

Both performance events were a celebration of the contribution pioneering female students have made to Australian society and beyond. They had a similar underlining message: that whilst a career in science is today a viable and exciting option for female students, at times the achievements of women in science in Australia have been/are undervalued. As well as a lack of acceptance by male lecturers and academic colleagues, in the nineteenth and early to mid-twentieth centuries female students had to battle misogynistic attitudes such as the perceived role women should perform in society, and whether they should be studying at university. Perhaps not surprisingly, as records at the University of Melbourne attest, many female students and alumni in the early 1900s joined, and even took a leading role in the women's suffrage movement (University of Melbourne, 2020).

When further analysing the data we had collected on the performances we realised that, as we later detail in our model, the ensemble or group-devised nature of the performances is not unlike a task many Australian secondary school drama students are asked to complete. Significantly, we noted that the style of performance is also akin to the 'non-naturalistic' or 'eclectic' style of theatre presentation that can often characterise secondary school ensemble or group-devised drama performances in Australia (Bird & Sallis, 2019). Whilst not labelled as such by its creators, the performance style of *A Journey Through Time* bore resemblance to the 'Poor Theatre' productions of Polish theatre director and theorist Jerzy Grotowski (2002) and that of English theatre director Peter Brook (1968). That is, the focus appeared to be on the technique and skill of the actor rather than on elaborate staging and technical effects. Consistent with this style of presentation, the set and property items (props) were kept to a minimum, with their function and meaning being transformed during the performance: the same prop could represent a variety of different objects depending on how the actor used it. The performance space and audience configuration were also consistent with the 'Poor Theatre' style. The theatre space, in this instance, was non-traditional, and it was simply lit. The audience surrounded the performance area and at times were interspersed within it.

Being an immersive theatre piece (Warren, 2017), the audience gathered at a preordained location and awaited instruction. The meeting point was the gothic entrance to a carpark under the South Lawn. In accordance with many immersive theatre productions, the number of audience members was kept to a manageable size. It soon became evident that this was because the spectators were to be drawn into the drama, rather than being passive observers. What the audience witnessed was a 'live exhibition' with the boundaries between actor and audience members almost non-existent; the relationship between actor and audience was intimate and dialogic. However, the 'guide' ensured that there was structure to the proceedings by instructing the audience members where to stand, when to interact with the characters and when to move on.

For both its original production and the reimagined version in 2018, *Ida* was a group-devised or ensemble performance which drew on a significant repository of research conducted by the ensemble members. Whilst both 'Ida' productions were musical comedies, they were nonetheless, for the most, part historically accurate and detailed some of the more notorious exploits of the 'Parlour' and the famous/infamous alumni who were its members. As Bonello et al. (2018) proffer in their text *Comedy and critical thought: Laughter as resistance*, audiences should not underestimate the power comedy has to subvert, mock, satirise critique and change thought. As we canvas in the teaching and learning model that follows, it would be a misapprehension to assume that a research-based theatre performance based on a seemingly serious topic such as women in science, necessarily has to be presented with a suitably unsmiling tone. As Bonello and her colleagues argue, the use of comedy or lighter moments within a dramatic performance can be just as moving and convincing. As author and public intellectual, Paul Goodman once said, "comedy is something that we can all share, no matter what language we speak or our background, it has the power to unite us all". In regard to performed research, as Bird et al. (2010) demonstrated most effectively in their production *Alice Hoy is not a building-Women in Academia*, a lighter touch to the writing and performing of such a play can still convey diligently conducted research and highly valuable research data and findings.

Teaching and Learning Model: 'Women in Science', Utilising Play-Making Techniques of Ensemble (or Group-Devised) Performance and Research-Based Theatre

What follows is a teaching and learning model we have developed for this chapter. In part the process is informed by our senior school drama textbooks *Acting Smart, Drama* (Bird & Sallis, 2019) and *Acting Smart, Theatre Studies* (Bird et al., 2019a). Significantly, the model we share below has been adapted for the non-drama classroom and for students and teachers who may be less familiar with play-making processes used in drama. Drama teachers are of course welcome to utilise it in their classrooms as well and may wish to show their performance to other members of the school community.

Below we propose a topic that could be given to a class in middle to senior secondary school. This can be adapted for the needs of a particular student cohort and for more junior classes if required. As is evident, the topic we have created for this task provides a context for the performance and suggests characters or character types which might inhabit the world of the play. Students in middle to upper secondary classes (e.g. Science, Humanities or English) are to be the devisers of the group-devised play. Whilst with ensemble performances it is customary that those who devise the piece also perform it, for this task it is possible that a class (for example a Science class) could devise the performance text and then others (e.g. students in a Drama class) could take the written text, rehearse and perform it. In Drama in schools, the teacher often divides the class into groups of between three and six students who each form an 'ensemble' (group) to devise their response to a set topic.

We suggest this grouping strategy could also work when devising a performance based on the topic below. Students could either form their own ensemble groups based on common responses to the given topic (see below) or be allocated groups by their teacher. Given the thematic focus of the performance (i.e. women in science) an appropriate gender balance across the groups may be beneficial in coeducational schools.

Topic for the Ensemble Performance

The topic below would be given to each of the ensemble groups in the class. Each group would then arrive at its own response to the topic. In this way each ensemble group will get to show its interpretation and will also be able to witness the responses of other groups in the class.

It is 2051 and time travel is finally possible due to the great Wormhole discovery of 2037. For International Women's Day, held on March 8th, 2051, a group of Australian women scientists from across the ages is being brought together for the first time. At the gathering there is one notable female scientist from each decade spanning the 1920s to the 2020s. The scientists have been brought together to share with each other their greatest achievements. However, whilst doing so, it becomes apparent that the road to their success was not easy and that they faced many obstacles such as studying Science at Secondary school, attaining their original university Science degree, gaining employment, and having their achievements officially recognised in what was, in those times, a male-dominated field. Their historic coming together highlights for the onlookers what had changed and what had remained the same over that one-hundred-year period. What also becomes apparent is why science is important in our lives and why there is a need for more students, of any gender, to study Science at school.

Intended Audience

When we work with our university students on a group devised play, we ask them to consider the extent to which they will involve their audience within the performance. For example, having one or more of the characters engaging in conversation with audience members, asking the audience their opinion about an issue raised in the play, inviting audience members to interact with the characters on stage. Akin to the *International Women's Day* performance we observed, students may wish to consider an immersive experience, where the audience members not only watch the performance but become a part of it. Alternatively, a more conventional or traditional staging may be preferred, where there is a clear delineation between actor and audience. However, in accordance with the tenets of research-based theatre, some form of audience interaction during or after the performance is encouraged.

In the development of a piece of research-based theatre, it is paramount that an intended or target audience is identified from the outset. Doing so informs aesthetic, thematic, contextual and content decisions that are made throughout all stages of the play-making process. It is suggested that for this project whilst peers, and parents and others within the school community will view the performance, a perceived target audience might be students at the point at which Science becomes an elective for them at the school. An aim therefore of the play-making project overall is to create a play which encourages students, females in particular, to consider choosing Science as a subject in their senior secondary school years. Given the constraints of a school timetable, it would be best if the repertoire of ensemble performances could be performed in a 90-min to 100-min block with each play not exceeding twenty minutes.

Step-By-Step Play-Making Process

Step 1: Negotiating, Collaborating and Brainstorming

Just like Scientists working on an experimental project, working on an ensemble or group-devised performance is a creative and a collaborative process. We recommend that once they have been shown the topic, each group starts by collectively brainstorming its initial response/overall concept. In accordance with a general principle of brainstorming, this early in the play-making process, the group members should be encouraged to be accepting of all ideas and not to reject or self-censor them.

Step 2: Research

Once each group has arrived at its initial responses to the topic, it is time to conduct the necessary associated research. Given the performed research style of the project, research will be an integral and important element of the play-making process. Not all students may be familiar with research techniques and the possibilities open to them. They may require assistance in regard to where and how to access information. For example, students can be directed to books such as *Amazing Australian Women* by Freeman and Beer (2018) and *Profiles, Australian Women Scientists* by Bharthal (1999). There are numerous relevant websites including the *Bright Sparcs–Teachers Guide*: *Australian Women Scientists* (University of Melbourne, 1997), *Women in STEMM, Australia* (https //womeninscienceaust.org) and *Supporting and encouraging women in Science* (University of Sydney, 2018). Students will also find a number of YouTube© and Vimeo© videos associated with the topic as well. An equitable process may be to have each member of the ensemble research a different facet of what is required and then prepare a verbal or written report for the other members of their group. Where possible, as well as interviews and TED talks that already exist online, it would be efficacious to have an Australian female Scientist guest speaker(s) talk to the students and be available for subsequent interview or questions (via email or social media). When deciding which real-life scientists the students want to research, one question for them to consider is what is meant by a 'woman in science'. For instance, is the woman in question someone who works/has worked as a scientist, or are they a teacher of science, or currently studying a Science degree? Will any male scientists also feature in the play and if so, for what purpose?

Ethical Considerations

An ethical question that can sometimes be raised by audience members in regard to research-based theatre, is whether or not the words spoken by the characters are quoted directly from the words of the research participants (Ackroyd & O'Toole, 2010). In our related work we have found that audience members often ask the extent to which the dialogue is fashioned from the actual words as spoken by the real-life people on whom the play is based. From an ethnodramatist's perspective, putting the words of the real-life person(s) in the mouth of the character(s) can help to add an authenticity to the performance; it is helpful to provide a direct link back to the research data on which the dramatic work is based. This principle is akin to that applied by social scientists when they write up their qualitative research reports. That is, they quote directly from research participants to help build a case for the validity of the findings being presented (Sallis, 2014). We recommend that when students in schools embark on an ethnodramatic project they too apply this principle. From experience, students enjoy the challenge this brings to their work and it enhances their engagement in the project (Sallis, 2010). However, such information may not always be available. If so, an evidence-based approximation or a guestimate of a real-life conversation may suffice. When commencing, the research students will benefit from knowing, when it comes to quoting from the real-life women scientists, what principle is to be applied and subsequently how to communicate the approach taken to the audience. Another important ethical principle to adopt is to ensure that the information selected from the research is balanced (and not slanted in a particular way purely for the purpose of dramatic impact). That is, where possible both sides of an argument or situation should be presented to avoid bias and misrepresentation. For example, in *Ida–The Musical*, not all of the male characters were unsympathetic to the women's cause. This resulted in a more nuanced view presented of what took place at the time.

Step 3: Preparing an Outline of the Plot of the Play

Once the overall concept for the performance and the bulk of the corresponding research into potential characters and events has taken place, the next key step is to determine the plot outline of the performance. Typically an outline of this kind might include: ideas for roles and characters; a possible plotline; the order of the scenes (which may be chronological or instead move forwards and backwards in time from a central point); the setting(s) to be depicted in the performance; a rationale for why the plotline will be suitable for the intended audience; possible themes to be explored, and what specific elements of the performance style will be in the performance (for example if the style is Brecht's Epic Theatre, will placards and direct audience address be used?). As previously mentioned, students may wish to adopt a comical/satirical performance style for their play, or mix comedy with a more serious tone.

Step 4: Improvising and Scripting

At this stage each ensemble group begins to improvise scenes, referring to its brainstorming responses and the 'Outline' it has prepared. When improvising at this stage, each group works on its own. In works of this kind it is not necessary to improvise the scenes in chronological order, especially if the group is confident of its plot structure for the performance. When improvising it can be useful to vary who plays which character in the scenes. Each group member will bring something different to the character when they portray them which will help them to develop the character and how it could be performed. There can be a tendency for students less familiar with drama to eschew (literally) standing up and improvising (they remain desk-bound and prefer to continue to write down their ideas than using improvisation to develop their play). Whilst some students may prefer to work in this way, it is in their best interests to have a go at improvising their scenes. If each group has its own space within the classroom to improvise, they are more likely to feel relaxed in doing so.

As scenes are improvised over and over, the group will start to gain a further insight into what their eventual performance may be like. At this point it is time for the group to re-assess their ideas, accepting some and rejecting others, which then leads to the scripting of the play. The nature of the resultant script may vary from group to group within the class. At one extreme, it may resemble a fully-formed script, complete with dialogue and stage directions. At the other, the script may be a record of each scene in order with a description of what will take place within it. For an ensemble performance, the former option is less common. One important aspect that students should realise is that the more dialogue they write, the more lines they have to learn. It is far more common that for an ensemble performance the learning of what to say and do comes from repeated improvising and, as in the next stage, rehearsal. However, from an ethical perspective, if the developing performance is incorporating quotations from the real-life persons, it is important that these are learnt by the performers and delivered accurately.

Step 5: Rehearsing–Acting, Directing, Editing and Refining

A key difference between the rehearsal and improvisation stages is that during rehearsals the performance starts to become locked down as final decisions are made about the order of the scenes, how characters will be portrayed, the nature and form of the interaction with the audience, and what stagecraft will be applied to accompany the acting (for example, costumes, props, use of multimedia, sound, lighting). With an ensemble performance the group needs to work together on all aspects of it, including its rehearsal and presentation. This may for example, include who will operate the sound, lighting and multimedia within the performance. If a video presentation or slideshow is to be incorporated (especially containing examples from the research phase such as photographs, video and historical documents) the group will need to determine who will be responsible for creating this content. As a point of interest, when promoting the 2019 *International Women's Day* events, including the *A Journey Through Time* performance, the University of Melbourne created an accompanying video (University of Melbourne, 2019).

In an ensemble performance, group members often share the directing of the piece. It can prove to be highly effective if each person in the group has a turn at standing outside a scene as it's being rehearsed to provide constructive feedback. However, this needs to be finely balanced otherwise a group member may spend "too much time directing at the expense of the development of their own character(s)" (Bird & Sallis, 2019, p. 40). When taking on the director role the student needs to revisit the original brief (topic) and see if it has been met. For example, are the Australian women scientists central to the performance? Does the research that was conducted about the scientists shine through or has any fictional content contained within the play unnecessarily overshadowed it? Is the aim of the performance sufficiently foregrounded, that is, to espouse the positives and benefits of studying Science, especially for female students? Rehearsal is not only about content, but also about the aesthetic look of the research-based performance as well. What refinements can be made to the way in which the story is being performed that can enhance the overall look and audience appeal of the performance?

Step 6: Presentation

The timing and the venue for the performance is something we recommend is determined well in advance to maximise its effectiveness. For example, will it be performed at a careers night at the school; will it form part of a subject-selection briefing for students and their parents; will it be held at the school as part of Science week? If the school has a performing arts theatre this may be used, however, for a more relaxed performance style, a multi-purpose classroom may be deemed more suitable.

One characteristic of research-based performances is that they are dialogic, meaning that the audience is given the opportunity to respond to the performance after, or sometimes during, its presentation (Denzin, 2003). Given the topic and research-based nature of each of the ensemble performances to be presented, it is important for each performing group to decide how it intends to interact with the audience. That is, how might it interact with the audience and how will it invite those watching the performance to respond to and reflect on what is being presented? For example, in Jane's Doctoral dissertation the actors who portrayed the female academic characters (and other characters) were 'hot-seated' at times during performance. That is, the characters were posed questions by the audience and the actors answered them in character (Bird, 2015). As part of his doctoral studies, Richard's research-based play (about the Drama classroom being a gendered space) was performed by students in a senior school Drama class. When presented, akin to Augusto Boal's Forum Theatre technique (1993), the play was performed twice. The first time was without interaction or interruption. The second time, members of the audience (students, teachers and others in the school community) were invited to stop the performance at any time and ask questions of the ethnodramatist (i.e. the playwright, Richard) and/or the performers before the play resumed once more.

Step 7: Reflection

Following the performance event, each ensemble group should be invited to reflect on both the devising process and their presentation. Were the aims and objectives met, and what is the evidence for this? Did audience members respond in a way that suggests the intention of the performance was realised? What have the ensemble group members and the audience learnt about Australian women scientists and the current status of science in Australia that perhaps they would not have known previously? Has this new knowledge changed their outlook on the role of women scientists and the role of science overall? What effect did the research-based ensemble performance have on achieving the aims of the project? What has been learnt about the potential of this form of performance for conveying content about science?

Conclusion

The 'steps' we have included in our Science/Drama teaching and learning model, and the order in which we recommend they be enacted, has come from our experience as drama teachers, researchers and ethnodramatists. Additionally, the ethical considerations we have highlighted are, we believe, important considerations when working on a research-based theatre project in a school setting.

Theatre is a powerful medium with which to convey content based on real people and events to audiences, including those in schools (Carter & Sallis, 2019; Sallis,

2014). Research-based theatre, or performed research is a particularly effective form in which to do this. There is a growing body of evidence that audiences respond positively to watching and responding to performances of this kind and that such plays can be an agent for change (Belliveau et al., 2020). This is often because audiences find they can empathise with, and relate to, the characters and situations depicted and take something from the play into their own lives. For example, *A Journey Through Time* generated a number of social media posts, especially from current female students studying science and medicine at the university, empathising with the events depicted about the female scientists in the play. For this reason, we argue that the teaching and learning model we have developed for this chapter has the potential to celebrate the achievements of Australian women in science and generate discussion about the topic in schools. Further, it may encourage students in secondary schools (including those in the class(es) who devise the ensemble plays), to consider the possibility of choosing science as a career and to further appreciate the contribution Australian women have made in the field of Science.

References

Ackroyd, J., & O'Toole, J. (2010). *Performing Research, tensions, triumphs and trade-offs of ethnodrama*. Trentham Books Limited.

Belliveau, G., & Lea, G. W. (2011). Research-based theatre in education. In S. Schonmann (Ed.), *Key concepts in theatre/drama education* (pp. 333–338). Sense Publishers.

Belliveau, G., & Lea, G. W. (2016). *Research based theatre, an artistic methodology*. Intellect Books.

Belliveau, G., Lea, G. W., & Westwood, M. (Eds.). (2020). *Contact!unload: Military veterans, trauma, and research-based theatre*. University of British Columbia Press.

Bharthal, R. S. (1999). *Profiles, Australian women scientists*. National Library of Australia.

Bird, J. (2015). *An interactive ethnographic performance: Ethnography, theatre and drama pedagogy for a professional learning context* (Doctoral dissertation). The University of Melbourne. Retrieved from https://minerva-access.unimelb.edu.au/bitstream/handle/11343/56513/Bird,%20J.%20Thesis%20to%20print.pdf?sequence=1.

Bird, J., Donelan, K., Sinclair, C., & Wales, P. (2010). Alice Hoy is not a building–Women in academia. In J. Ackroyd & J. O'Toole (Eds.), *Performing research: Tensions, triumphs and trade-offs of ethnodrama* (1st ed., pp. 81–103). Trentham Books.

Bird, J., & Sallis, R. (2019). *Acting smart, drama, Version 8*. Acting Smart Publishing Group.

Bird, J., Sallis, R., & Bailey, M. (2019). *Acting smart, theatre studies, version 8*. Acting Smart Publishing Group.

Bird, J., McConville, K., & Sallis, R. (2019). 'The Greatest Love of All': An ethnodrama celebrating the continuing and evolving culture of the drama teacher. *NJ*, *43*(2), 69–82. https://doi.org/10.1080/14452294.2019.1704975.

Boal, A. (1993). *Theatre of the oppressed*. Theatre Communications Group.

Bonello, K., Giappone, R., Francis F., & MacKenzie, I. (Eds.). (2018). *Comedy and critical thought: Laughter as resistance*. Rowan and Littlefield International.

Brook, P. (1968). *The empty space*. Penguin.

Carter, C., & Sallis, R. (2019). Investigating the role of drama in two enabling courses in Australia. *Applied Theatre Research, 7*(1), 79–93.

Dasgupta, N., & Stout, J. G. (2014). Girls and women in science, technology, engineering, and mathematics: STEMing the tide and broadening participation in STEM careers. *Policy Insights from the Behavioral and Brain Sciences, 1*(1), 21–29. https://doi.org/10.1177/2372732214549471

Denzin, N. K. (2003). *Performance ethnography: Critical pedagogy and the politics of culture.* Sage.

Freeman, P., & Beer, S. (2018). *Amazing Australian women: Twelve women who shaped history.* Hachette Australia.

Grotowski, J. (2002). *Towards a poor theatre.* Routledge.

Hatton, C. (2020). Imagining the possible: Using drama for gender equality in schools. In Á. H. Ragnarsdóttir & H. S. Björnsson (Eds.), *Drama in education: Exploring key research concepts and effective strategies* (pp. 128–146). Routledge.

Hatton, C., & Sallis, R. (2019). A research tango in Three Moves: Gendering the drama research space. In P. Duffy, C. Hatton, & R. Sallis (Eds.), *Drama research methods, provocations of practice* (pp. 57–76). Brill/Sense.

Kang, J., Hense, J., Scheersoi, A., & Keinonen, T. (2019). Gender study on the relationships between science interest and future career perspectives. *International Journal of Science Education, 41*(1), 80–101. https://doi.org/10.1080/09500693.2018.1534021

Kalive, P. (2018). *Director's notes.* Retrieved from umsu.unimelb.edu.au/ida.

O'Toole, J., Stinson, M., & Moore, T. (2009). *Drama and curriculum—a giant at the door.* Springer.

Saldaña, J. (2005). *Ethnodrama: An anthology of reality theatre.* AltaMira Press.

Saldaña, J. (2011). *Ethnotheatre: Research from page to stage.* Left Coast Press.

Sallis, R. (2010). *The drama of boys, an ethnographic study and performance* (Doctoral dissertation). University of Melbourne. Retrieved from https://minerva-access.unimelb.edu.au/handle/11343/36929.

Sallis, R. (2014). Ethnographic performance: A change agent for drama teaching and learning. *Research in Drama Education: THe Journal of Applied Theatre and Performance, 19*(3), 313–325. https://doi.org/10.1080/13569783.2014.928011

Salmon, A. (2015). A complex formula: Girls and women in science, technology, engineering and mathematics in Asia. *UNESCO Bangkok.*

Schön, D. A. (1984). *The reflective practitioner, how professionals think in action.* Basic Books.

Sinclair, C., Jeanneret, N., O'Toole, J., & Hunter, M. (Eds.). (2017). *Education in the arts* (3rd ed.). Oxford University Press.

Soans, R. (2012). Robin Soans. In W. Hammond & D. Steward (Eds.), *Verbatim: Contemporary documentary theatre* (pp. 15–44). Oberon Books.

University of Melbourne. (1997). *Bright Sparcs–Teachers guide: Australian women scientists.* Retrieved from https://www.asap.unimelb.edu.au/bsparcs/guides/t_teachers.htm.

University of Melbourne. (2019, March 8). *A journey through time* [Streaming video]. Retrieved from https://www.youtube.com/watch?v=4KW0fPyeqBc.

University of Melbourne. (2020). *Women in the archives.* Retrieved from https://archives.unimelb.edu.au/resources/subject_guides/women-in-the-archives/women-at-the-university-of-melbourne.

University of Sydney. (2018). *Supporting and encouraging women in Science.* Retrieved from https://www.sydney.edu.au/science/about/women-in-science.html.

Warren, J. (2017). *Creating worlds, how to make immersive theatre.* Nick Hearn Books.

Yin, R. (2013). *Case study research, design and methods.* Sage Publications.

Richard Johnson Sallis (B.Ed., M.Ed., Ph.D.) is a senior lecturer in drama/theatre education in the Artistic and Creative team in the Melbourne Graduate School of Education (MGSE) at The University of Melbourne. He is joint editor of JACE (Journal of Artistic and Creative Education). Richard is a life member of Drama Victoria, a former President of Drama Australia, and the past Director of Publications for IDEA (the International Drama/theatre in Education Association). His areas of interest include: Arts Based Research (ABR), research-based theatre (ethnodrama/performed research), diversity and inclusion in schools, gender/sexualities and schooling, drama/theatre education, the education of pre-service teachers, transnational education, and cultural influences on teaching and learning.

Jane Bird (B.Ed., M.Ed., Ph.D.) is a senior lecturer in drama/theatre education in the Artistic and Creative team in the Melbourne Graduate School of Education (MGSE) at The University of Melbourne. She specialises in the artistic, embodied and collaborative qualities of teaching and learning in and through drama and theatre. Jane researches a range of applications for drama pedagogy across disciplines and contexts. She has developed multiple pieces of research-based theatre and written about the construction processes and aesthetic nature of performance ethnography.

Science, Drama and the Aesthetic

Russell Tytler and Vaughan Prain

Abstract In this chapter we review the variety of themes opened up by the authors in this book and situate these themes within a semiotic aesthetic theoretical framing drawing on the work of the pragmatist semiotician Charles Sanders Peirce. We trace the historical divergence and interactions between art and science to argue that both disciplines create, reason about, and communicate knowledge and meaning through the inventive development and use of sign systems, infused with disciplinary aesthetics/values. We claim that this focus on meaning-making through sign systems allows us to see more clearly multiple opportunities for the symbiotic relationship between drama and science advocated by many chapter authors and by the editors in their Chapter "Sparking Learning in Science and Drama: Setting the Scene" scene setting. Drawing on Peirce, we outline both (a) the interrelations between experiences of phenomena, sign systems and meaning-making and (b) the inevitable entwining of feeling and meaning in this process. We argue that a Peircean theoretical lens explains and warrants the value of drama's embodied, narrative semiotic resources in supporting students' construction of meaning in science classrooms, and their appreciation of dramatic forms in opening up meanings that have conceptual and human dimensions. We claim that this enrichment of focus and methods is a key to students developing a positive aesthetic sense of scientific ways of reasoning and acting in the world. Conversely, drama's aesthetic perspective and semiotic resources are well placed to explore and interrogate themes, successes, and contemporary issues raised by science.

Keywords Science · Drama · Pragmatist semiotics · Aesthetics · Narrative · Meaning-making

---

R. Tytler (✉)
School of Education, Deakin University, Burwood, VIC, Australia
e-mail: russell.tytler@deakin.edu.au

V. Prain
Faculty of Arts & Education, Deakin University, Waurn Ponds, Australia
e-mail: vaughan.prain@deakin.edu.au

P. J White et al. (eds.), *Science and Drama: Contemporary and Creative Approaches to Teaching and Learning*, https://doi.org/10.1007/978-3-030-84401-1_16

Introduction

This book's advocacy of diverse ways that drama and science can productively interact raises interesting and fundamental questions about the nature of disciplines and their corresponding epistemic practices, and how these might interrelate. In particular, questions about drama-science relationships raise wider issues such as: the relations between the arts, including drama, and science; the nature of each and the ways in which they diverge and overlap historically; and intersections between the distinct ways they make and value meaning in the world.

In this chapter we consider these issues from a pragmatist perspective on semiotics and aesthetics, drawing particularly on the work of Peirce (1907/1998). We claim that aesthetics in science and drama in school entails students developing both an active taste for and appreciation of the goals, foci, multi-modal meaning-making resources, processes, and methods in each discipline, and that this aesthetic sense is therefore fundamental to both how and what is learnt in each subject. We consider that there is a complementarity in the distinctive but overlapping sign systems by which meanings are generated, experienced, and valued in both disciplines to support learning. We further consider how the aesthetic invitation of drama as embodied personalised felt experience can enrich student science conceptual learning, and how engaging with the diverse aspects of science can also enrich the focus in drama. We point to emerging research findings about the value of engaging with the aesthetic dimension of sign systems in learning science (Lehrer & Schauble, 2012) and how using an expanded range of sign systems afforded by the arts can enrich this learning (Tytler et al., 2020a, b).

The chapters as a set provide diverse perspectives on drama, reflecting various traditions within this discipline concerning different purposes, methods and performance genres. The drama activities described range from role-plays of physical phenomena (Paige, Brown, O'Keeffe, & Garrett, Chap. "Dramatising the S and M in STEM"; Davis, Chap. "Dramatic and Undramatic Emotional Energy: Creating Aesthetic and Emotive Learning Experiences in Science Classrooms"), gaming including quiz style games (Paige, Brown, O'Keeffe & Garrett, Chap. "Dramatising the S and M in STEM"; Siklander & Harmoninen, Chap. "Ice Age is Approaching: Triggering University Students' Interest and Engagement in Gamified Role-Play Outdoor Playful Learning Activities"), role-plays of social controversies related to science (Hannigan & Ferguson, Chap. "Art-Science Education in the Anthropocene: Embodied Metaphor with Puppets and Performance"; Henderson & King, Chap. ""This is the Funniest Lesson": The Production of Positive Emotions During Role-Play in the Middle Years Science Classroom"; Myers, Chap. "New Eden: Mediating Pre-service Teachers' Conceptions of Education for Sustainable Development"; Raphael & White, Chap. "Transdisciplinarity: Science and Drama Education Developing Teachers for the Future"), historical or contemporary practice dramas using a 'Mantle of Expert' approach (Baskerville & Anderson, Chap. "Responding to Climate Change: Developing Primary Children's Capability to Engage with Science

Through Drama"; Clark-Fookes & Henderson, Chap. "Mutuality and Inter-relativity of Drama and Science"; Swanson, Chap. "Does Being Positioned in an Expert Scientist Role Enhance 11–13 Year-Old Students' Perceptions of Themselves as Scientists?"), or dramatising pedagogical moves such as demonstrations of scientific processes and concepts (Davis, Chap. "The Treatment of Dr. Lister: Investigating and Revisiting a Famous Primary Science-Based Drama"; McGregor, Chap. "Stories from History: More Authentic Ways of Thinking Through Acting and Talking About Science"; van Cuylenburg, Chap. "The Science Drama Project: Meaning in the Middle"). Similarly, aligned with these different meanings foregrounded and promoted by these dramatic forms, we also see a range of perspectives on learning science that extend well beyond a traditional account of learning as focused on abstracted truths about the world. These chapters include an emphasis, for instance, on appreciation of the role of scientists (Chaps. "Mutuality and Inter-relativity of Drama and Science", "The Science Drama Project: Meaning in the Middle", "The Treatment of Dr. Lister: Investigating and Revisiting a Famous Primary Science-Based Drama", "Responding to Climate Change: Developing Primary Children's Capability to Engage with Science Through Drama", "New Eden: Mediating Pre-service Teachers' Conceptions of Education for Sustainable Development", "Stories from History: More Authentic Ways of Thinking Through Acting and Talking About Science"), of the nature of science (Chaps. "Mutuality and Inter-relativity of Drama and Science", "The Treatment of Dr. Lister: Investigating and Revisiting a Famous Primary Science-Based Drama", "Dramatising the S and M in STEM", "Responding to Climate Change: Developing Primary Children's Capability to Engage with Science Through Drama", "Does Being Positioned in an Expert Scientist Role Enhance 11–13 Year-Old Students' Perceptions of Themselves as Scientists?"), of ethical entailments of scientific practices (Chaps. "The Science Drama Project: Meaning in the Middle", "Transdisciplinarity: Science and Drama Education Developing Teachers for the Future", ""This is the Funniest Lesson": The Production of Positive Emotions During Role-Play in the Middle Years Science Classroom"), of positive emotions associated with science learning (Chaps. ""This is the Funniest Lesson": The Production of Positive Emotions During Role-Play in the Middle Years Science Classroom", "Dramatic and Undramatic Emotional Energy: Creating Aesthetic and Emotive Learning Experiences in Science Classrooms"), including interest, and identity and aspirations towards science positioning and science futures (Chaps. "Responding to Climate Change: Developing Primary Children's Capability to Engage with Science Through Drama", "New Eden: Mediating Pre-service Teachers' Conceptions of Education for Sustainable Development", "Ice Age is Approaching: Triggering University Students' Interest and Engagement in Gamified Role-Play Outdoor Playful Learning Activities", "Does Being Positioned in an Expert Scientist Role Enhance 11–13 Year-Old Students' Perceptions of Themselves as Scientists?", "Australian Women in Science: A Model for a Research-Based Theatre Project in Secondary School Classrooms"). We argue that this list is consistent with widening curriculum perspectives on the purposes of a science education, which renders this drama-science focus particularly timely.

These questions of the interrelations between yet distinctiveness of art and science as disciplines have relevance for the way we think of the two disciplines of drama and science interacting in a curricular sense. In Raphael, White and van Cuylenburg's overview in Chap. "Sparking Learning in Science and Drama: Setting the Scene", the editors take up our model (Tytler et al., 2019), developed for interdisciplinary STEM, that highlights the temporal nature of the interaction over different scales. This implies that at any, or at least at most points in time, one can discern whether the activity can be identified as involving drama, or science practices. At the macro scale, this accords attention to the building of practice competencies in each discipline over time as an important feature of the interdisciplinarity, a point made in many chapters about the mutuality of the relationship, and emphasised particularly in Chap. "Sparking Learning in Science and Drama: Setting the Scene" as a principle of mutual symbiosis. The nature of this mutualism is the point we wish to examine more closely in this chapter. We suspect that the relations are different between art/drama and science than they are between the STEM subjects, because the epistemological foundations are more distinct, and may lead to different ways they intertwine. Separating mathematics and science aspects of an activity may be more straightforward, for instance, than separating the drama aspect of an embodied role-play from resultant science learning. We pursue this question of distinctiveness/integration from the perspective of the ways that the arts/drama and science make meaning in the world. We do this, drawing particularly on the aesthetic and semiotic perspectives of pragmatist writers, particularly Charles Sanders Peirce, but also Dewey (1897/1972), and more recent scholars such as Lemke (1990, 2004) and Wickman (2006). The chapter structure is built around the questions:

1. What are the historical distinctions and interrelations between art and science?
2. What are the semiotic means by which each discipline creates meaning?
3. How can an aesthetic perspective broaden our understandings of learning in science, and in drama, and the way that each can support richer learning in the other?
4. What research is needed to move this area forward?

Historical Divergence and Overlap of the Arts and Science

As noted in Chap. "Sparking Learning in Science and Drama: Setting the Scene", there is a history of attempts to define and legitimate the arts and science as oppositional, but also a rival history of attempts to argue for unification, symbiosis, and to propose overlap in methods, foci and learning outcomes. Jones and Galison (2013) point out that the binary case entails polar reputational opposites: "soft versus hard, intuitive versus analytical, inductive versus deductive, visual versus logical, random versus systematic, autonomous versus collaborative, and, like all binaries, at some level, female versus male" (p. 2). These researchers also claimed that "the binary production of knowledge (the bifurcation of practices) was equally simple: art invented, science discovered" (Jones & Galison, 2013, p. 2). However, they and contributors to this book demonstrate in many ways the reductive inadequacy of this binary case. While the disciplines have divergent purposes, varied epistemologies, contrastive contexts, histories, valued practices and tools, and different peer review processes, both disciplines also have histories of overlapping practices for creating valued meanings. Contemporary perspectives make it clear that science is both found and made, and that science and drama create new realities rather than simply 'unveil' pre-existing truths. Science and the arts entail the invention of models/representations (Lehrer & Schauble, 2006; Nersessian, 2008) that create new accounts of phenomena (Gooding, 2004, 2006), and both disciplines use old and new sign systems to make and share new meanings. There is abundant evidence that artistic and scientific sensibilities are not distinct, and numerous examples, some quoted in the chapters in this volume, of individuals who combine artistic and scientific sensibilities and competencies such as Leonardo da Vinci and Albert Einstein. There are persistent claims made that the creative impulse in science is similar to that in art, even though the forms of evidence to support new knowledge differ in the two areas.

In the next section we call on socio-semiotic and pragmatist perspectives to unpack how knowledge in science, as with drama, is constructed, and apprehended. In this, we draw on Peirce's semiotics, and his aesthetics, to explore how art and science, and drama and science, fundamentally intersect in the production of meaning in both disciplines.

Socio-semiotics, Meaning and Feeling in Science and Art

In thinking about the construction of meaning in science and in drama, and the learning process in the school representations of these disciplines, we draw on the semiotics of the pragmatist scholar Charles Sanders Peirce. Peirce's semiotics constitute an account of how we generate meaning through creating and reasoning through sign systems, and this necessarily applies to both science and art. Figure 1 shows Peirce's (1907/1998) triad through which meaning is created through the iterative interplay between a referent or object, a sign representing that object, and the interpretant as the meaning that is made of this relationship. According to this perspective, semiosis, or the creation of meaning, is a recursive, material-conceptual process involving systems of signs in different modes.

Science and drama exist as highly evolved cultural practices with elaborate valued material and symbolic resources (sign systems) for seeking and making meanings through representing possible, speculative and modeled experiences and worlds. Scientists integrate linguistic, mathematical, visual and embodied/actional modes in tandem with material manipulation to make warranted claims. This claim-making is matched in drama by multimodal resources used to explore, critique, and represent felt experiences and worlds through integrating linguistic, actional/interactional, gestural, visual, temporal, spatial signs. Both disciplines overlap in their need

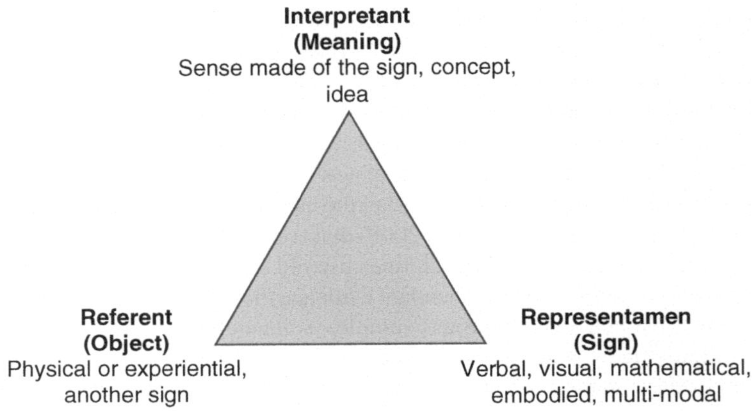

Fig. 1 Peirce's meaning making triad

to integrate modes to create, review and critique modeled and possible experiences, but drama's aesthetic particularly invites personal felt engagement in the meaning-making, whether as participant or spectator.

Aesthetics and Meaning

Crucial to learning any school subject are students' spontaneous and learnt felt responses to what they are expected to notice, use and do. This is variously described in the literature as attitudes (for example to topics in science or to science as a subject), emotions (pleasure, boredom, anxiety, confusion) in response to what is going on in the classroom, or motivation and interest. These characterisations tend to regard affect as extraneous to conceptual learning, a sort of by-product of classroom processes. However, from a pragmatist semiotic perspective, the affective and conceptual are inextricably linked. In Peirce's semiotics, our initial meaning occurs in a form that Peirce (1907/1998) called the *emotional interpretant* that colours and is inextricably bound with the ongoing process of semiosis/meaning-making. Peirce accorded aesthetics a prime role in his philosophy of signs and meaning, emphasising the crucial role of evaluative feelings in the meanings conferred on phenomena and experiences.

To Peirce, questions of truth are intimately bound up with informed action, and thus with the purposes and values directing conduct. Ultimately this determines individuals' judgments about their preferences; what they like or not (Tejera, 1994). Dewey (1897), also from a pragmatist perspective, regarded the conceptual and the aesthetic as 'continuous'; one cannot exist without the other. Dewey and later scholars (Girod & Wong, 2002; Girod et al., 2003; Hobbs, 2012) refer to aesthetic experience as being significant and often formative for deeper learning, involving a close link between the conceptual and aesthetic. Lemke (2015), drawing on Peirce, regarded feelings as inseparable from meaning-making, entailing aesthetic processes that are "distributed, situated, context-dependent, active and culture-specific". He argued that "feeling and meaning are coeval, coevolved, functionally complementary, co-determined, and co-determinative" (Lemke, 2015, p. 602). The distinctive nature of the aesthetic construct, compared to more generic formulations of affect, lies in its association with specific objects, purposes and outcomes associated with particular subjects, whether these be material objects (such as a worm; Bloom, 1992), experiences (such as being out in the environment, or taking part in a role-play), conceptual constructs (such as the elegance or power of theories), or practices (such as particular drama forms).

Knowledge Construction, Aesthetics and Learning in Science

There is increasing agreement amongst scholars that scientific discovery processes entail imaginative and communal processes of new tools and models, and the translation of these through a process of representational redescription. Gooding (2004, 2006) analysed Michael Faraday's notebooks to identify the complex imaginative, informal and visual reasoning through which knowledge of electromagnetic fields was generated. Our own research in science education is premised on the principle that school science should better represent the epistemic practices by which knowledge is built in science (Prain & Tytler, 2012), and also foreground the aesthetic dimensions to this learning (Ferguson et al., 2021; Wickman, 2006). We see learning in science classrooms as necessarily involving multimodal, informal and creative abductive reasoning that is infused with aesthetic choices and values (see Chap. "Art-Science Education in the Anthropocene: Embodied Metaphor with Puppets and Performance").

There is a growing literature demonstrating the crucial role of aesthetic judgments in learning and knowing in science and in learning science. Wickman (2006) demonstrated the aesthetic commitments of scientists in developing new knowledge, and of students negotiating understanding and developing interest in scientific ways of looking at the world (Anderhag et al., 2015). Jakobson and Wickman (2008) demonstrated that teachers' aesthetic focus enticed students into grappling with conceptual learning. Ferguson et al. (2021) described how the process of learning data modeling processes involved the shift from an 'art' aesthetic to a disciplinary aesthetic of appreciation of a data set and ways of productively exploring this mathematically.

A scientific aesthetic may relate to many aspects of the scientific enterprise, such as the elegance or power of scientific ideas, the precision and/or complexity of method, the determination of scientists in search of knowledge or solutions, the values and ethical entailments of scientific activity, or the complexity of interactions of science with other forms of knowing. We have advocated the development of a broader repertoire of representations (Prain & Tytler, 2012) extending to drama or art. New approaches to science learning and thinking about the purposes of a science education pose new challenges but provide opportunities for new insights into how students' aesthetic responses and their take-up can productively influence science learning and engagement with science (Campbell, 2018; de Mesa, 2018; Jakobson & Wickman, 2015; Tytler, Prain & Hannigan, 2020a, b). We have worked with interdisciplinary art-science learning sequences to show the distinct but overlapping and mutually reinforcing disciplinary aesthetics of these subjects (Hannigan, Wickman, Ferguson, Prain & Tytler, 2021; Tytler, Prain & Hannigan, 2020a, b).

These 'within science' aspects of aesthetics and learning are all represented in these chapters, and the cases demonstrate clearly that drama foregrounds these aesthetic foci in meaning-seeking and–sharing. In this sense, drama becomes an important way to open up and offer the possibilities of science as encouraging identity-formation as students learn to think and act as a scientist; or play in figured worlds (Urrieta, 2007) that open up imaginative spaces for the shaping of aesthetic commitments.

In theorising this representation-focused approach we have developed a model of representation construction affordances (Prain & Tytler, 2012) through which we align classroom practices with those of the scientific community. We argue that each representation's contribution to learning and meaning is always partial and approximate and can be understood in terms of its affordances (Gibson, 1979) which operate as productive constraints on thinking. For instance, a drawing achieves its affordance through its visual and spatial specificity, such as when students need to decide on spatial arrangement and movement of particles when interpreting evaporation in terms of submicroscopic processes, or on the specific alignment of the earth, earth orbit and earth location and angle when explaining the angle of the sun in the sky (Tytler, Prain, Aranda, Ferguson & Gorur, 2020a, b). By implication, a broader range of relevant sign systems can support success in learning.

In our approach, learners generate, share and consolidate meaning through linking across multiple sign systems, with each sign adding a particular affordance that interacts with and enriches the interpretation of other signs and material objects. We argue that drama, in the many ways illustrated in this book, offers extended sign systems that enrich how and what students can experience, value and learn in science.

Representational Aspects of Drama in Relation to Science

The entailments of this semiotically informed perspective on learning are pertinent to the question of drama-science interdisciplinarity. These relate to the nature of reasoning in meaning-making as encompassing perceptual, metaphorical elements and to the narrative structure of much reasoning. This also includes the role of embodied representations in establishing meaning in science.

First, from this socio-semiotic perspective, reasoning is multifaceted and not constrained by formal logic. Research into learning through representation construction, re-description across modes and coordination, emphasises the role of visualisation and spatial reasoning. Further these perceptual, informal reasoning processes very often encompass metaphor and analogy that open up complex relations between signs, such as water references in explaining electric circuits, or constructs such as niche in ecology, or the heart as a pump. Further, in solving problems and generating explanations, abduction is crucially important as an imaginative reasoning process. These perspectives open up the possibility of drama as a productive tool for meaning-making through a variety of semiotic resources and practices (see van Cuylenburg's exploration of dramatic forms in relation to scientific aspects of human experience, Chap. "The Science Drama Project: Meaning in the Middle"), beyond that implied by a view of learning in science as requiring formal reasoning processes around abstract linguistically expressed concepts.

Second, sign systems also include embodied representations of scientific phenomena. The body is a powerful semiotic resource for meaning making, for instance through gesture and haptic experience, and through bodily positioning. We see this operating in role-plays where for instance a role-play of atoms in a solid as it heats affords a productive constraint on students' attention to particle spacing and movement during change of state processes (Prain & Tytler, 2012). Such embodied representations offer perceptual insights. The centrality of the body is supported by more recent neuroscience findings that imagining the arrangements of submicro-scopic particles involves a complex perceptual operation involving positioning and resizing (Gibbins, 2013), and empirical evidence supports the support of learning through active participation and experience (Kontra et al., 2015).

Third, the introduction of drama into the science classroom often includes a narrative structure or context. This is illustrated in many forms in the chapters in this book. The psychologist Bruner (1985, 1991) makes the point that narrative is a fundamental resource for constructing and explaining complex influences on sequences of events and cause/effect relationships. Science reputedly deals with abstracted universals largely independent of context, yet for Bruner, narrative creation and review can be a central means to create and interpret complex phenomena and change. In all these chapters in this book, we can see this thread of narrative adding richness to the way students are invited to make meaning of and engage with the concepts and practices of science.

We also argue that many scientific concepts can be viewed as compressed narratives in terms of their origins in the history of science, and as nominalised accounts of processes. Meaning-making in science involves reasoning that is sequential but can be grounded in narrative, even if this reasoning is more formally codified through abstracted visual and symbolic sign systems than is often the case in drama. Thus, explaining the movement of electrons in an electric circuit (see Chap. "Dramatising the S and M in STEM") involves complex causal and categoric interrelations involving multiple actors, that lends itself readily to bodily interpretation involving an unfolding narrative. The nominalisation process characteristic of science ('digestion' stands in for a complex temporal process) hides this underlying narrative structure which drama is well positioned to unpack and clarify, and embodied improvisation of the type described in a number of chapters can provide strong support for representational re-description across modes, such as between visual diagrams and symbolic and textual representations. An example of this narrative structure can be seen in the effectiveness of animation construction processes such as slowmation (Hoban & Nielsen, 2014) in supporting student understanding. White et al. (2020) showed how students constructing an animation of digestive processes grappled afresh with the categorical and temporal relations between entities in a process unspecified in the written text and static image representations in the textbook.

These potentialities of dramatic enactment are evident in the chapters of this book and indicate a variety of different dimensions to thinking and acting scientifically and encouraging aesthetic engagement. In some cases these involve the embodied creation of a causal explanatory narrative to unpack categorical and causal relations to explain the science (such as exploring the operation of a bee hive through story: Clark-Fookes & Henderson, Chap. "Mutuality and Inter-relativity of Drama and Science"; or constructing role-plays of gear operation, or electric circuit relations: Paige, Brown, O'Keeffe, & Garrett, Chap. "Dramatising the S and M in STEM"). In other cases the focus is on other dimensions of science understandings, such as the personal and ethical issues around stem cell technologies (Raphael & White, Chap. "Transdisciplinarity: Science and Drama Education Developing Teachers for the Future"), transplant recipients (Henderson & King, Chap. ""This is the Funniest Lesson": The Production of Positive Emotions During Role-Play in the Middle Years Science Classroom"), scientific perspectives on death or human action (van Cuylenburg, Chap. "The Science Drama Project: Meaning in the Middle"), understandings of intersections between gender and science (Sallis & Bird, Chap. "Australian Women in Science: A Model for a Research-Based Theatre Project in Secondary School Classrooms"), or understanding of microbiology through dramatic historical recreation (such as dramatising the work of Lister: Davis, Chap. "The Treatment of Dr. Lister: Investigating and Revisiting a Famous Primary Science-Based Drama"). Human narratives focusing on science settings and practices were featured in Baskerville & Anderson's (Chap. "Responding to Climate Change: Developing Primary Children's Capability to Engage with Science Through Drama") account of students in the role of scientific advisors developing environmental policy, in Myers' (Chap. "New Eden: Mediating Pre-service Teachers' Conceptions of Education for Sustainable Development") account of exploring developmental issues in a protected forest, and in recreating investigation of the sinking of the Wahine (Swanson, Chap. "Does Being Positioned in an Expert Scientist Role Enhance 11–13 Year-Old Students' Perceptions of Themselves as Scientists?"). We see in this range of foci the multiple dimensions of science understanding that can be made more meaningful by the narrative structures provided by drama.

Research Directions

There is a need, as pointed out in Raphael, White, and van Cuylenberg's Chap. "Sparking Learning in Science and Drama: Setting the Scene" overview, to develop a more nuanced perspective on drama pedagogies in relation to science, and to theorise their interrelationship. We suggest that the wider framing of the agendas in both science and drama offered by the focus on the creation of meaning (see, for instance, van Cuylenburg, Chap. "The Science Drama Project: Meaning in the Middle", or Swanson, Chap. "Does Being Positioned in an Expert Scientist Role Enhance 11–13 Year-Old Students' Perceptions of Themselves as Scientists?"), and of a semiotically infused aesthetic, could be productive in further exploring a variety of drama-science interactions and providing a coherence to this agenda.

The key advantage of an aesthetic focus, compared with a focus on emotions or engagement, lies in its sharp association with the disciplinary objects that are the proper focus of an education. Thus, rather than asking the questions–were students happy? Were they active?–the question 'how do they respond to the scientific object or process' is more in keeping with a science learning intent. This explicit aesthetic focus on developing appreciation of and absorption in the objects and processes of scientific ways of thinking and working, or of the role and processes of drama, can be seen in a number of chapters, such as: the photographs in Davis' Chap. Dramatic and Undramatic Emotional Energy: Creating Aesthetic and Emotive Learning Experiences in Science Classrooms" " showing students' clear focus on the flame; the questions of interest in applying science to a problem flagged by Pirkko Siklander and Sari Harmoinen (Chap. "Ice Age is Approaching: Triggering University Students' Interest and Engagement in Gamified Role-Play Outdoor Playful Learning Activities"); or in concern with the structure of drama in White and Raphael's STEM cell exploration (Chap. "Transdisciplinarity: Science and Drama Education Developing Teachers for the Future"); or in students' appreciation of dramatic inquiry described by Myers (Chap. "New Eden: Mediating Pre-service Teachers' Conceptions of Education for Sustainable Development").

If we are to better understand the aesthetic entailments of these drama-science interactions, we need to develop methodological strategies for how to elicit and track spontaneous and longer-term aesthetic engagement of students and teachers with scientific phenomena and practices (Prain, Wickman, & Ferguson, under review). Such an agenda is consistent with the increasing interest in identity as an appropriate focus for students' engagement with thinking and working scientifically (Barton et al., 2013; Tytler, 2014) or in drama (Roy & Ladwig, 2015) and with acceptance of the wider remit of an education to include transdisciplinary competencies.

Conclusion

In the context of this book, we argue that the languages of drama, and of science, and the ways we as educators speak of these fields, can be productively broadened to focus explicitly on aesthetic dimensions, such as appreciation, interest, and developing a taste, sensibility and disposition for scientific ways of working, for scientific cultures of discovery, and for the processes and cultures of drama. The construct of aesthetics offers key insights into the multiple ways that drama and science can interact and a richer account of the range and nature of meaning-making processes in science and drama.

In terms of the contribution of drama to science education, we argue that drama-science combinations are well placed to pursue some of the key directions of a contemporary science education. These include the focus on identity (role-playing scientists' thinking and persona), on decision making, on recognising the role of scientific knowledge in relation to other knowledge forms, societal implications and ethical positions, on citizenship competencies, and on scientific dispositions that include engagement with the epistemic bases of the discipline. Drama affords multiple opportunities to anchor, concretise and review what otherwise would remain abstracted and remote. The chapters in this book all foreground the human aspects of science, and as such, the space opened up by drama for imaginative projection into scientific experiences and reasoning processes.

In this chapter we have focused mainly on how aesthetic engagement with science can be enriched through incorporating drama's resources for aesthetic meaning-making through what is felt and known in immersion and enactment. Students as both participants and spectators can learn through embodied representations of science phenomena and through various dramatic processes, such as Mantle of the Expert, process drama, and gamification. However, conversely, drama can also be enriched by applying its semiotic resources to exploring and engaging with the many themes, achievements and issues that science practices and values raise. These include: (a) the human aspects of scientific discovery processes and decision-making based in science, (b) the integrity of scientific explanatory and problem-solving processes in societal contexts, (c) the unforeseen effects of science on humanity, and (d) personal meaning and values in relation to scientific phenomena and explanations. In these reciprocal ways, the semiotic resources of each discipline and their associated aesthetics can mutually enhance what students feel, experience and come to understand and value in and across each subject. The chapters in this book show that when the sign systems of drama and science are combined, resultant transdisciplinary meanings (the human, ethical, narrative turn in science, the principles and practices in drama that can make sense of science as a major human undertaking) deepen student learning. These transdisciplinary effects, so routinely sought and practised now in the world, should inform what is learnt across subjects in school.

References

Anderhag, H., Hamza, K. M., & Wickman, P.-O. (2015). What can a teacher do to support students' interest in science? A study of the constitution of taste in a science classroom. *Research in Science Education, 45*(5), 749–784.

Barton, A., Kang, H., Tan, E., O'Neill, T., Bautista-Guerra, J., & Brecklin, C. (2013). Crafting a future in science: Tracing middle school girls' identity work over time and space. *American Educational Research Journal, 50*(1), 37–75.

Bloom, J. D. (1992). The development of scientific knowledge in elementary school children: A context of meaning perspective. *Science Education, 76*, 399–413.

Bruner, J. (1985). Narrative and paradigmatic modes of thought. *Learning and Teaching the Ways of Knowing, 84*, 97–115.

Bruner, J. (1991). The narrative construction of reality. *Critical Enquiry, 18*(1), 1–21.
Campbell, C. (2018). In search of our beginnings: Locating 'Firstness' in arts education in the service of advocacy. *International Journal of Education & The Arts, 19* (3).
de Mesa, J. A. L. (2018). Peirce and aesthetic education. *Journal of Philosophy of Education, 52*(2), 246–261.
Dewey, J. (1897/1972). *The aesthetic element in education*, reprinted in 1972, John Dewey: The Early Work, 1882–1898 (Vol. 5). Southern Illinois University.
Dewey, J. (1934/1987). *Art as experience*, reprinted in 1987, John Dewey: The Late Work, 1925–1953 (Vol. 10). Southern Illinois University.
Ferguson, J., Tytler, R., & White, P. (2021). The role of aesthetics in the teaching and learning of data modelling. *International Journal of Science Education (special Issue on Aesthetics in Science Education).* https://doi.org/10.1080/09500693.2021.1875514
Gibbins, I. (2013). A feeling for the image: Hands, body and visualisation of the invisible. In C. Kennedy & M. Rosengren (Eds), *SPECTRA: Images and data in art/science Proceedings from the symposium SPECTRA 2012* (pp. 129–134). Australian Network of Art and Technology (ANAT).
Gibson, J. (1979). *The ecological approach to visual perception.* Houghton Mifflin.
Girod, M., Rau, C., & Schepige, A. (2003). Appreciating the beauty of science ideas: Teaching for aesthetic understanding. *Science Education, 87*(4), 574–587.
Girod, M., & Wong, D. (2002). An aesthetic (Deweyan) perspective on science learning: Case studies of three fourth graders. *The Elementary School Journal, 102*(3), 199–224.
Gooding, D. (2004). Visualization, inference and explanation in the sciences. In G. Malcolm (Ed.), *Studies in multidisciplinarity* (Vol. 2, pp. 1–25). Elsevier.
Gooding, D. (2006). From phenomenology to field theory: Faraday's visual reasoning. *Perspectives on Science, 14*(1), 40–65.
Hannigan, S., Wickman, P-O., Ferguson, J., Prain, P., & Tytler, R. (2021). The role of aesthetics in learning science in an art-science lesson. *International Journal of Science Education (Special issue on Aesthetics in Science Education).* https://www.tandfonline.com/doi/full/10.1080/09500693.2021.1909773
Hoban, G., & Nielsen, W. (2014). Generating science discussions through creating a narrated stop-motion animation: The affordances of Slowmation. *Teaching and Teacher Education, 42*, 68–78.
Hobbs, L. (2012). Examining the aesthetic dimensions of teaching: Relationships between teacher knowledge, identity and passion. *Teaching and Teacher Education, 28*(5), 718–727.
Jakobson, B., & Wickman, P.-O. (2008). The roles of aesthetic experience in elementary school science. *Research in Science Education, 38*(1), 45–65.
Jakobson, B., & Wickman, P.-O. (2015). What difference does art make in science? A comparative study of meaning-making at elementary school. *Interchange, 46*(4), 323–343.
Jones, C., & Galison, P. (2013). Introduction to picturing science, producing art. In C. Jones & P. Galison (Eds.), *Picturing science, producing art* (pp. 1–26). Routledge.
Kontra, C., Lyons, D. J., Fischer, S. M., & Beilock, S. L. (2015). Physical experience enhances science learning. *Psychological Science.* https://doi.org/10.1177/0956797615569355
Lehrer, R., & Schauble, L. (2006). Cultivating model-based reasoning in science education. In K. Sawyer (Ed.), *Cambridge handbook of the learning sciences* (pp. 371–388). Cambridge University Press.
Lehrer, R., & Schauble, L. (2012). Seeding evolutionary thinking by engaging children in modeling its foundations. *Science Education, 96*(4), 701–724.
Lemke, J. (2004). The literacies of science. In E. W. Saul (Ed.), *Crossing borders in literacy and science instruction: Perspectives on theory and practice* (pp. 33–47). International Reading Association/National Science Teachers Association.
Lemke, J. L. (1990). *Talking science: Language, learning and values.* Greenwood Publishing Group.
Lemke, J. (2015). Feeling and meaning: A unitary bio-semiotic account. In P. P. Trifonas (Ed.), *International handbook of semiotics* (pp. 589–616). Springer.

Nersessian, N. (2008). Model-based reasoning in scientific practice. In R. Duschl & R. Grandy (Eds.), *Teaching scientific inquiry: Recommendations for research and implementation* (pp. 57–79). Sense Publishers.

Peirce, C. S. (1907/1998). Pragmatism. In N. Houser, A. De Tienne, J. R. Eller, C. L. Clark, A. C. Lewis & D. B. Davis (Eds.), *The essential Peirce—Selected philosophical writings—Volume 2 (1893–1913)* (pp. 398–433). Indiana University Press.

Prain, V., Wickman, P-O., & Ferguson, J. (under review). Addressing methodological challenges n researh on aesthetic dimensions to classroom science inquiry. *International Journal of Science Education (Special issue on Aesthetics in Science Education).*

Prain, V., & Tytler, R. (2012). Learning through constructing representations in science: A framework of representational construction affordances. *International Journal of Science Education, 34*(17), 2751–2773.

Roy, D., & Ladwig, J. (2015). Identity and the arts: Using drama and masks as a pedagogical tool to support identity development in adolescence. *Creative Education, 6*(10), 907.

Tejera, V. (1994). The primacy of the aesthetic in Peirce and classic American philosophy. In H. Parret (Ed.) *Peirce and value theory: On Peircian ethics and aesthetics* (pp. 85–97). John Benjamins.

Tytler, R. (2014). Attitudes, identity, and aspirations toward science. In N. G. Lederman & S. K. Abell (Eds.), *Handbook of research on science education* (Vol. 2, pp. 82–103). Routledge.

Tytler, R., Prain, V., Aranda, G., Ferguson, J., & Gorur, R. (2020). Drawing to reason and learn in science. *Journal of Research in Science Teaching, 57*, 209–231. https://doi.org/10.1002/tea.21590

Tytler, R., Prain, V., & Hannigan, S. (2020). Rethinking the languages of science and how they are learnt. *Research in Science Education.* https://doi.org/10.1007/s11165-020-09952-8

Tytler, R., Prain, V., & Hobbs, L. (2019). Re-conceptualising interdisciplinarity in STEM through a temporal model. *Research in Science Education.* https://doi.org/10.1007/s11165-019-09872-2

Urrieta, L. (2007). Figured worlds and education: An introduction to the special issue. *The Urban Review, 39*(2), 107–116.

White, P. J., Tytler, R., & Nielsen, W. (2020). Animation construction as cross-modal translation in senior biology. In L. Unsworth (Ed.), *Learning from animations in science education* (pp. 209–228). Springer.

Wickman, P.-O. (2006). *Aesthetic experience in science education: Learning and meaning-making as situated talk and action.* Lawrence Erlbaum Associates.

Russell Tytler (B.Sc (Hons), M.Sc, M.Ed, Ph.D. FASSA) is Alfred Deakin Professor and Chair in Science Education at Deakin University, Melbourne. He has researched and written extensively on student learning and reasoning in science. His interest in the role of representation as a multimodal language for reasoning in science extends to pedagogy and teacher learning. He researches and writes on student engagement with science and mathematics, the role of affect, aesthetics and identity in learning, school-community partnerships, and STEM curriculum policy and practice. His current interest is in interdisciplinarity leading to critical and creative reasoning. He is widely published and has been chief investigator on a range of Australian Research Council and other research projects.

Vaughan Prain (BA (Hons), Dip Ed, M.Ed, Ph.D.) is a Professor in Science Interdisciplinary Education Research at Deakin University. He has an extensive research record in the role of writing for learning in science, and more recently in how this mode relates to other modes, such as visual, mathematical and embodied modes in reasoning about, constructing understanding of, scientific concepts and processes. His current research focus is on how learning in science can be enhanced by incorporating strategies and approaches used in other subjects, including visual arts.

The manufacturer's authorised representative in the EU is Springer Nature Customer Service Centre GmbH, Europaplatz 3, 69115 Heidelberg, Germany. If you have any concerns regarding our products, please contact ProductSafety@springernature.com

Printed and bound by CPI Group (UK) Ltd, Croydon, CR0 4YY
30/09/2026
02225508-0008